T0205988

Undergraduate Lecture Notes in Physics

More information about this series at http://www.springer.com/series/8917

Undergraduate Lecture Notes in Physics (ULNP) publishes authoritative texts covering topics throughout pure and applied physics. Each title in the series is suitable as a basis for undergraduate instruction, typically containing practice problems, worked examples, chapter summaries, and suggestions for further reading.

ULNP titles must provide at least one of the following:

- An exceptionally clear and concise treatment of a standard undergraduate subject.
- A solid undergraduate-level introduction to a graduate, advanced, or nonstandard subject.
- A novel perspective or an unusual approach to teaching a subject.

ULNP especially encourages new, original, and idiosyncratic approaches to physics teaching at the undergraduate level.

The purpose of ULNP is to provide intriguing, absorbing books that will continue to be the reader's preferred reference throughout their academic career.

Series editors

Neil Ashby
Professor Emeritus, University of Colorado Boulder, CO, USA

William Brantley
Professor, Furman University, Greenville, SC, USA

Matthew Deady
Professor, Bard College, Annandale, NY, USA

Michael Fowler
Professor, University of Virginia, Charlottesville, VA, USA

Morton Hjorth-Jensen
Professor, University of Oslo, Norway

Michael Inglis
Professor, SUNY Suffolk County Community College, Selden, NY, USA

Heinz Klose
Professor Emeritus, Humboldt University Berlin, Germany

Helmy Sherif
Professor, University of Alberta, Edmonton, AB, Canada

Todd Keene Timberlake • J. Wilson Mixon, Jr.

Classical Mechanics
with *Maxima*

 Springer

Todd Keene Timberlake
Berry College
Mount Berry, GA, USA

J. Wilson Mixon, Jr.
Berry College
Mount Berry, GA, USA

ISSN 2192-4791   ISSN 2192-4805   (electronic)
Undergraduate Lecture Notes in Physics
ISBN 978-1-4939-3206-1   ISBN 978-1-4939-3207-8   (eBook)
DOI 10.1007/978-1-4939-3207-8

Library of Congress Control Number: 2015950472

Springer New York Heidelberg Dordrecht London

Printed on acid-free paper

Springer Science+Business Media LLC New York is part of Springer Science+Business Media (www.springer.com)

Preface

Why Did We Write *Classical Mechanics with Maxima*? Computational tools have become a vital part of doing physics in the twenty-first century. However, the college physics curriculum has been slow to include instruction in computation. Some departments may not include computation in the physics curriculum because they do not have the staffing, or the space in their curriculum, to add an additional course in computational physics. This problem is quite common for small physics departments. One way around the problem is to incorporate computational instruction into an existing, standard physics course. However, faculty teaching those courses may not feel that they have the resources to guide them in carrying out this task. *Classical Mechanics with Maxima* is intended to help solve this problem. It is meant to supplement a standard textbook for an undergraduate classical mechanics course. The aim of this book is to provide an opportunity for students to learn computation, using the *Maxima* computer algebra system, while they are also learning the standard topics of classical mechanics.

Why Did We Write <u>*Classical Mechanics* with *Maxima*</u>? One reason we chose to focus on classical mechanics is that the study of classical mechanics does not require knowledge of physics beyond what most students obtain in their first semester of introductory physics. Because of these minimal prerequisites, computation can be incorporated into a classical mechanics course that students take early in their undergraduate careers, which allows those students to use computation throughout their undergraduate education. Also, classical mechanics is more intuitive and provides more opportunities for visualization (particle trajectories, etc.) than other areas of physics. Finally, introducing computation into classical mechanics allows students to see the limitations of an analytical approach to the subject. Specifically, computation allows students to explore the exciting field of chaotic dynamics.

Why Did We Write *Classical Mechanics with <u>Maxima</u>*? Why did we choose to use a computer algebra system (CAS) rather than a standard programming language like Java or a scripting language like Python? The main reason for this choice is that computer algebra systems offer both symbolic and numerical computing capabilities. Thus, computer algebra systems can be useful for students even when they are solving a problem analytically. Most CASs have built-in visualization tools (for plots, etc.). In addition, the learning curve for a CAS is not as steep as for a traditional programming language and CASs are generally easier to use. Therefore, the time from conception to finished calculation is often smaller when using a CAS versus a programming language. Computer algebra systems have become a standard part of the physicist's toolkit, in part because they allow the user to focus on physics rather than on programming.

Many powerful CASs are available. Perhaps, the most popular are *Mathematica*™ and *Maple*™.[1] So why did we choose to use *Maxima* for this book? By far the most important reason was cost. *Maxima* is free for everyone, which means that students can easily install a copy on their own computer. They are not restricted to using the software in a computer lab. Likewise, students will always have access to *Maxima* no matter where their career takes them. While *Mathematica*™ and *Maple*™ offer reasonably priced student licenses, those licenses are not transportable once students graduate and the standard licenses for those programs are quite expensive. We also like *Maxima* because it is open-source and is maintained by an active community of developers. It is easy to install and use on any operating system, and its feature set, while not as extensive as that of *Mathematica*™, is more than sufficient for undergraduate physics instruction. Most CASs are similar enough that users who learn *Maxima* should have an easy time transitioning to another CAS should they need or want to do so.

Why Did We Write *Classical Mechanics with Maxima*? It all began when Wilson discovered *Maxima* and began using it as a tool for teaching Economics. Meanwhile, Todd had been teaching classical mechanics using a different CAS. Todd became frustrated with the high cost of the CAS license and was looking to move to a cheaper (preferably free and open-source) alternative for a new course he was developing on computational physics. Todd had just about settled on a combination of *Maxima* and *Easy Java Simulations* for his new course when he was approached by Wilson about writing a book similar to Ronald Greene's *Classical Mechanics with Maple*™ but using *Maxima* instead. Todd had been teaching classical mechanics using a CAS for years, and Wilson had the expertise to help Todd with the transition to *Maxima*, so writing the book seemed like a no-brainer. Although we were originally inspired by Greene's book, we took our book in a different direction, with much more emphasis on numerical computation and algorithms and how these methods can be used to illustrate important ideas in physics. The result, we hope, is a book that blends both physics and computation together in ways that are mutually complementary.

---

[1]*MATLAB*™ is also very popular, especially with engineers, but it is not a true CAS, although it now includes some symbolic computing capability.

How Should You Read *Classical Mechanics with Maxima*? This book should not be read like a regular physics textbook (much less like a novel!). The *Maxima* code to accompany each chapter can be obtained from our website:

sites.berry.edu/ttimberlake/cm_maxima/.

The website also has links to a variety of other resources for using *Maxima*. We strongly recommend that you read the book while also having a computer available to run the corresponding *Maxima* code. Evaluate the code within *Maxima* as you read about it in the book. Modify the code and play around with it to get a feel for what it does. Work the exercises at the end of the chapter (this goes for ANY physics textbook!). Many standard classical mechanics textbooks include computational problems, and *Maxima* can be used to help solve analytical problems as well, so you can apply your *Maxima* skills to problems from these books as well.[2] We have included a larger number of exercises on those topics (such as Chaps. 5 and 6) that are not typically included in the standard textbooks. Once you have worked your way through this book, you can apply the computational tools you have learned to more advanced topics in classical mechanics and other areas of physics. You can even extend your knowledge of *Maxima* using the built-in Help feature or various online resources, or expand your computational toolkit by learning new computational tools, many of which are free and open-source just like *Maxima*.[3]

We close this preface by acknowledging those who have helped this book come to life. Todd thanks the former professors and collaborators who taught him how to do computational physics, particularly Matthew Choptuik, Mario Belloni, and Wolfgang Christian. Todd also thanks the students from his classical mechanics and computational physics courses who gave him valuable feedback on much of the content of this book. Wilson thanks his wife Barbara for suggesting that he learn the use of a computer algebra system, for reading all of *Microeconomic Theory and Computation* which he coauthored, and for drawing on her knowledge of physical chemistry in helping him with his work on this volume.

Mount Berry, GA, USA

Todd Keene Timberlake
J. Wilson Mixon, Jr.

---

[2]Some examples of standard textbooks for an undergraduate classical mechanics course are *Classical Mechanics* by Taylor, *Classical Dynamics of Particles and Systems* by Thornton and Marion, and *Analytical Mechanics* by Fowles and Cassiday.

[3]We particularly recommend *Easy Java Simulations* for building interactive computer simulations.

Contents

Chapter 1
Basic Newtonian Physics with *Maxima*

1.1 Introduction to *Maxima*

Classical mechanics is the branch of physics that deals with the motion of objects subject to forces and constraints. Classical mechanics has existed as a well-defined subject since the publication of Newton's *Principia Mathematica Philosophiae Naturalis* in 1687. Since Newton, a few important new concepts have been introduced into the subject (such as energy and its conservation), but most of the developments in classical mechanics since the seventeenth century have consisted of new mathematical techniques for solving classical mechanics problems. The twentieth century saw the creation of an entirely new tool for addressing these kinds of problems: the digital computer.

Computers have affected many areas of physics, and classical mechanics is no exception. Computers have enabled physicists to study areas of classical mechanics that are entirely new, as well as some that lay long-neglected. These areas involve problems that are too difficult, or even impossible, to solve by hand. Computers offer the possibility of generating numerical solutions to these problems, thus breaking down a long-standing barrier. Computers also provide for the graphical representations of mathematical expressions and numerical data, helping physicists gain an intuitive visual understanding of classical mechanics problems and their solutions.

1.1.1 Computer Algebra Systems

With the development of computer algebra systems (CASs) that can perform symbolic mathematical manipulations, computers make it easier to solve the problems that can be solved by hand. Computer programs that can perform symbolic

© Todd Keene Timberlake & J. Wilson Mixon, Jr. 2016
T.K. Timberlake, J.W. Mixon, *Classical Mechanics with Maxima*, Undergraduate
Lecture Notes in Physics, DOI 10.1007/978-1-4939-3207-8_1

mathematical operations, carry out numerical computations, and generate graphical displays of formulas are now an indispensable tool in the study of classical mechanics. The open-source CAS *Maxima* can perform all of these operations.

This book focuses on *Maxima* as it applies to the study of classical mechanics. Much of the material that is developed here, however, applies beyond this context.[1] One of the attractions of software like *Maxima* is that one need not become an instant expert. Rather, learning a few basics and then applying them to a set of problems facilitates expanding the range of inquiry and the associated range of *Maxima*'s capabilities.

This chapter introduces *Maxima* as it applies to symbolic analysis. The material introduced in this chapter is used and extended throughout the remainder of the text. It is important to work through this material, trying out the examples in *Maxima*, before moving on to the rest of the material. This practice provides the background for understanding the examples that appear throughout the text.

This text does not detail the use of *Maxima* on any particular computer operating system. *Maxima* was developed to operate on a Linux platform, but is easily installed and used on both Windows and Macintosh operating systems, and versions exist for the Android and iOS operating systems as well.[2] The material developed here can be applied in any of these environments.

The remainder of this chapter introduces *Maxima* and the user interface *wxMaxima*.[3] It provides an overview of how *Maxima* can serve as a powerful calculator. This chapter also illustrates how *Maxima* can be used to review basic physics topics. This review includes an introduction to *Maxima*'s treatment of vectors, the solution of equations, calculus, the solution of ordinary differential equations (ODEs), root finding, and plotting.

1.1.2   Installing Maxima

Maxima can be installed on Windows, Mac, or Linux systems. To install *Maxima* you can visit the official *Maxima* site, http://maxima.sourceforge.net/. In this book we present *wxMaxima*, which provides a convenient user interface for the *Maxima* computer algebra system. You can download *wxMaxima* from http://wxmaxima. sourceforge.net/. Click the Download tab at the top of the page and then follow

---

[1] Also, *Maxima*'s similarity to other, proprietary software ensures that lessons learned in using this CAS will be useful even if the analyst's career involves the use of other software.

[2] Limited versions of *Maxima* are available on Android through the *Maxima on Android* app, and on iOS through the *Sage Math* app.

[3] This interface is not the only one available to *Maxima* users, but it is the one used throughout this text.

the instructions to download and install *wxMaxima* for your system. The *wxMaxima* package includes the complete *Maxima* installation, as well as required plotting software and fonts. *Maxima* and *wxMaxima* are updated frequently, so it is a good idea to check back regularly for new versions.

1.2 Interacting with *Maxima*

Maxima is primarily an interactive tool into which the user types and enters commands, causing the computer to respond to those commands immediately, though commands may be entered in a batch mode. *Maxima* displays a prompt, usually a `->` sign or an input prompt like this: `%i1`, when it awaits input. *Maxima* commands are typically algebraic expressions, assignments, or function definitions. *Maxima* is case-sensitive; it is important to pay attention to this fact when issuing *Maxima* commands.

Each piece of input must be terminated by a semicolon (;) or a dollar sign ($) to let *Maxima* know that it has reached the end of the command. A semicolon causes *Maxima* to print the result of executing the command, whereas a dollar sign suppresses that output but keeps the result in *Maxima*'s memory. A long command can be extended over several lines by hitting the Enter key to start a new line. *Maxima* tries to execute the command only when it encounters a `;` or a `$`, not at a line's end. In the *wxMaxima* interface, an input cell may contain any number of commands. These batches of commands are implemented when the cursor is placed anywhere within the cell and either a Shift Enter or a Control Enter combination is typed.

The following are simple examples of inputs and *Maxima* responses. Note the use of `*` for multiplication, `^` for raising to a power, and `!` for the factorial. Also, notice the numbering convention for input and output. The prompt for the first input is (`%i1`), and the output that results from executing the first command is output (`%o1`). Until this session is ended, the output (`%o1`) remains in *Maxima's* memory and can be recalled for use in further analysis.[4]

Observe that the second input spans two lines. *Maxima* input ignores lines. A single command can span any number of lines. Likewise, a single line can hold any number of commands. Three different inputs appear on the line labeled (`%i4`). Each of these inputs is given its own separate input number: %i4 for the first, %i5 for the second, and %i6 for the third. Observe also that input %i7 recalls two quantities

---

[4]The user can remove any or all of the entries that are stored by using the explicit `kill` command, which is used and discussed below.

that have been stored in Maxima's memory with assigned names %i1 and %i6. (The default is for *wxMaxima* to produce input and input labels in typewriter font and output and output labels in Roman font.)

```
(%i1)  4*5
(%o1)   20

(%i2)  1 + 2 + 3 + 4 + 5 +
6 + 7 + 8 ;
(%o2)   36

(%i3)  2*x - 5 + 3.5*x;
(%o3)   5.5x − 5

(%i4)  -2^4;  (-2)^4;  5!;
(%o4)   −16
(%o5)   16
(%o6)   120

(%i7)  %o1 + %o6;
(%o7)   140
```

Maxima does simple arithmetic both numerically, as in 4*5 returning 20, and symbolically, as in adding $2x$ and $3.5x$ to get $5.5x$. In addition to the four primary arithmetic operations $+$, $-$, $*$ and $/$, *Maxima* can exponentiate ($\char`\^$) and take factorials (!). Other operations can be performed with function calls.

If a command is one that *Maxima* does not recognize (*e.g.*, a misspelling of a command name or a capital letter that should have been lowercase, or a function not yet defined), *Maxima* usually returns the input as it was entered, with no explanation.

Maxima uses the mathematically correct order of arithmetic operations to evaluate an expression. Parentheses can be used to arrange a different order of operations, as in input %i5 above.

Often we need to refer to the result of the most recently completed operation. *Maxima* allows us to refer to the result of the last evaluated expression with a percentage sign (%). Expressions resulting in syntax errors do not count. The % reference in *Maxima* is to the most recently executed command, not necessarily the one that appears directly above the command being executed.[5] Thus, the user must keep track of the command sequence when using this feature. An example that makes use of the previous output is shown below.

```
(%i8)  (x + 3)^2; % - 5;
(%o8)   (x + 3)²
(%o9)   (x + 3)² − 5
```

---

[5]This is an important feature of *Maxima*. Although commands may be organized *spatially* within a *wxMaxima* notebook, information is stored in *Maxima*'s memory *chronologically*. We will see several examples of this throughout the book.

1.2.1 The wxMaxima *Screen*

The basic input of *Maxima* commands is the same for any *Maxima* user interface. In this book we assume that the reader is using the *wxMaxima* interface and therefore it will be useful to introduce some features that are particular to that interface. The *wxMaxima* screen shown in Fig. 1.1 consists of five parts. First, the title bar identifies the file being used, if a session has been saved as a named file. Next, a series of menus provides a relatively easy way of implementing most of *Maxima*'s important commands. A set of tutorials that provide an overview of how to use the interface, while demonstrating much about *Maxima*'s capabilities is at http://andrejv.github.io/wxmaxima/help.html.

The third row in the *wxMaxima* screen contains a set of commonly used icons. The first opens a new *wxMaxima* session. The second icon opens a previously created *wxMaxima* session. The third icon saves the current session. The fourth icon prints the current session, showing both input and output. The fifth icon allows for configuration of *wxMaxima*. The next two icons copy or cut a selection. The eighth icon pastes material from the clipboard. The ninth icon interrupts the execution of a command, which is useful if an infinite loop has been entered inadvertently. The next two icons control animation (not addressed in this text).

The fourth part of the screen is a window that contains input and output. Figure 1.1 shows two commands in an input group and the output that results from executing them. Also, a comment is included. All comments must be bracketed as indicated: /* ...text ...*/. A comment cannot be the last entry in an input group.

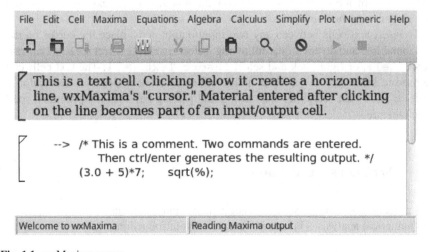

Fig. 1.1 wxMaxima screen

Clicking above or below the cell that contains those commands causes a horizontal line to appear. This is the *wxMaxima* cursor. Once the cursor appears, one can type input. Once the input group is complete, typing a combination of either the Shift key or the Control key and the Enter key submits the input to *Maxima* and the resulting output is returned. The result of one series of entries appears below.

```
(%i10) 3*4;  1 + 2 + 3 + 4 + 5;
       3*(x - 5)^2;
       w : (x + 3)^2; % - 5; w;
(%o10)  12
(%o11)  15
```
$$(\%o12) \quad 3\,(x-5)^2$$
$$(\%o13) \quad (x+3)^2$$
$$(\%o14) \quad (x+3)^2 - 5$$
$$(\%o15) \quad (x+3)^2$$

The use of input groups has a number of advantages, but it can cause some difficulty in matching input and output. Therefore, the number of commands entered into an input group should be chosen judiciously.

Most of the *Maxima* input and output discussed in this book will be presented in the form shown above. However, there are situations in which it is more convenient to present input or output within a paragraph of text. For example, the command 1 + 2 + 3 + 4 + 5 produces the output **15**. Input is represented in typewriter text. Output is presented in **bold**, using the standard font. When *Maxima* commands are displayed separately from the text, input is indicated by (%i), and output is indicated by (%o). Input and output numbers are suppressed. Either input or output can consist of more than one item. All inputs and outputs are gathered in cells in the *wxMaxima* workbook that accompanies the relevant text material, so you can see how the material in the text is produced and experiment with the commands.

To insert text cells between any two input/output cells, place the cursor on the horizontal line and do any of the following: select ctrl-1, select Cell/Insert Text Cell, or right-click and select Insert Text Cell. The text cells can be used to keep track of variables and relationships. The *wxMaxima* text editor is basic. It allows for searching and cutting and pasting but does not provide a spell check feature.

1.3   *Maxima* as a Calculator

The preceding section introduces basic features that Maxima offers for symbolic and numerical analysis. This section extends that introduction.

1.3.1 Data Types

Maxima deals with several kinds of numbers. Integers, rational numbers, irrational numbers, and floating-point numbers are fundamental. In addition, the program can manipulate complex numbers, whose real and imaginary parts can be any of the fundamental types. Except for floating-point numbers (i.e., numbers containing a decimal point), numbers are stored in *Maxima* as exact values. The *Maxima* default is to store floating-point values with an accuracy of 16 decimal points. This default can be changed, using the floating-point precision (`fprec`) command.

Whenever possible, *Maxima* tries to return exact answers for all calculations. Consider the evaluation of the natural logarithm of 10 and of the fraction 352/1200 using the list of commands below. The result is a list of values.

(%i) **[log(10), 352/1200];**   (%o)[log (10) , $\frac{22}{75}$].

Embedding this list of commands within a `float` command instructs *Maxima* to output floating-point representations of the values, which it does to the default 16 decimal places.

(%i) **float([log(10),352/1200]);**
(%o)[**2.302585092994046,   0.29333333333333**]

The `fpprintprec` (floating-point precision) command can be used to reduce the default number of decimal places. The command `fpprintprec : 5$`, placed before the preceding command, would result in the list [**2.3026,0.29333**]. *Maxima* retains these values at the default 16-place accuracy. To reset the number of digits reported, use the command `fpprintprec:0$`. The value of `fpprintprec` cannot be 1.

1.3.2 Mathematical Functions

Maxima knows about many mathematical functions, including how to evaluate them for specific argument values and how to manipulate them symbolically, to differentiate and integrate, to apply identities, etc. Some of the functions that are commonly used in physics are illustrated below. In addition, the inverse trigonometric and hyperbolic functions are available with the corresponding names appended by the character a: `asin`, `asech`, etc.

First consider these four built-in functions: finding an absolute value, extracting a square root, raising e ($\approx$ 2.718) to a power, and extracting the natural logarithm of a number. These operations are illustrated below using the number -44 as the argument. The first input below binds the name a to the number -44. The second input consists of a list of commands that generates the output list. In the first two listed items, *Maxima* reports exact answers. In the third, it reports a floating-point representation, as instructed. The fourth output item is an exact representation. In the fifth item *Maxima* evaluates log(−44) as a floating-point number (the result is a complex number, because we are taking the logarithm of a negative number).

```
(%i) a:-44$ [abs(a),sqrt(a),float(exp(a)),
     log(a), float(log(a))];
```
$$(\%o)\ [44, 2\ \sqrt{11}\,i,\ 7.78113\ 10^{-20},\ \log(-44),\ 3.1416\,i + 3.7842].$$

Maxima offers the expected complement of trigonometric functions, such as the three shown below. *Maxima* reports $\sin(\theta)$, $\cos(\theta)$, and $\tan(\theta)$ for $\theta = \pi/3$. Note that the letter π is indicated by %pi; "pi" alone would result in the letter but no value.

```
(%i)   theta : %pi/3$
       [x1,x2,x3]:[sin(theta),cos(theta),tan(theta)];
```
$$(\%o)\ [\tfrac{\sqrt{3}}{2}, \tfrac{1}{2}, \sqrt{3}].$$

The list below reports the arcsin, arccos, and arctan for the results above. These inverse trigonometric functions should just return the original argument used to evaluate the corresponding trigonometric function, so the result in each case should be the original angle ($\pi/3$). The *Maxima* calculations give the expected results.

```
(%i)  [ asin(x1), acos(x2), atan(x3) ];
```
$$(\%o)\ [\tfrac{\pi}{3}, \tfrac{\pi}{3}, \tfrac{\pi}{3}]$$

Commands can be embedded inside other commands. The first command below is equivalent to the two commands that comprise the second input line. *Maxima* works from the inside outward.

```
(%i)  asin( sin(5*%pi/3) );
      sin(5*%pi/3); asin(%);
```
$$(\%o)\ -\tfrac{\pi}{3} \quad -\tfrac{\sqrt{3}}{2} \quad -\tfrac{\pi}{3}$$

Maxima can also evaluate hyperbolic functions. For example, the commands below yield the list of values for the hyperbolic sine, cosine, and tangent of 0.5. Note that *Maxima* returns floating-point (decimal) output when it is given floating-point input.

```
(%i)  b:0.5$    [y1,y2,y3]:[sinh(b),cosh(b),tanh(b)];
      [asinh(y1),acosh(y2),atanh(y3) ];
```
$$(\%o)\ [0.5211, 1.1276, 0.46212] \quad [0.5, 0.5, 0.5]$$

Maxima can evaluate a variety of special functions. One example, the "Bessel function of the first kind," is illustrated with the two commands below. The first command controls the nature of the output. The second command results in the output in the second line.

```
(%i)  besselexpand:true$        bessel_j(3/2, z);
```
$$(\%o)\ \frac{\sqrt{2}\,\sqrt{z}\left(\frac{\sin(z)}{z^2} - \frac{\cos(z)}{z}\right)}{\sqrt{\pi}}$$

The list of commands below returns values of the Gaussian Error Function. *Maxima* can also report the Complementary Error Function, the Imaginary Error Function, and the Generalized Error Function.

(%i) [erf(0),erf(0.5),erf(2.0)];   (%o) $[0, 0.5205, 0.99532]$

Maxima can also evaluate the Gamma Function, Γ, which extends the factorial function to the real and complex values of n.

(%i) [gamma(4), gamma(4.5), gamma(1.0 + %i)];
(%o) $[6,\ 11.632,\ 0.49802 - 0.15495 * \%i]$

Maxima contains many other mathematical functions, as well as commands that allow us to manipulate expressions. Some appear later in this chapter. The *wxMaxima* menus provide a well-organized listing of commands, and working through these menus is a valuable exercise. For a complete listing of commands, refer to the *Maxima Manual*.

1.4   1D Kinematics: Variables and Functions

We now apply *Maxima* to some real physics. We begin with a review of basic Newtonian physics that is typically covered in an introductory course. Our first topic is one-dimensional kinematics. For example, we know that the position of an object moving with constant acceleration a is given by

$$x(t) = x_0 + v_0 t + \frac{1}{2}at^2, \tag{1.1}$$

where x_0 is the object's position at $t = 0$ and v_0 is the object's velocity at $t = 0$.

To use this result in *Maxima* we define a function that gives us x as a function of t. There are two ways we might choose to define this function. The first way is to assign an expression for calculating x to a *variable*. A variable in *Maxima* is just a symbol that can be assigned a particular value. The value of a variable can be a number, but it can also be an expression (or a list, or a matrix, etc.). If we want the variable *pos* to represent the position of our object at time t, then we can assign the appropriate expression to that variable. *Maxima* returns the expression that is assigned we name *pos* unless the command ends with a dollar sign.

(%i) pos:x0+v0*t+(1/2)*a*t^2;   (%o) $x0 + t\,v0 + \frac{at^2}{2}$

Note the use of the colon for assigning a value to a variable. Now if we want to evaluate the position of the object at a certain time, we can ask *Maxima* to evaluate the variable *pos* while substituting a particular value for t. For example, the position at $t = 5$ s is shown below. The subst (substitute) command replaces t with the value 5.

(%i) subst(t=5, pos);   (%o) $x0 + 5\,v0 + \frac{25\,a}{2}$

Note that *pos* is not a function. It does not have an argument. All we have done above is to evaluate the expression that is assigned to *pos* using a particular value of *t*. We could just as easily evaluate it for a particular value of *a*, or $v0$, etc. If we wish to know the position of the object at $t = 5$ s when $a = -9.8$ m/s^2, $x_0 = 0$, and $v_0 = 32$ m/s, we can make the necessary substitutions. When more than one substitution is to be made, a list of the substitutions is required.

```
(%i) subst([t=5, a=-9.8, x0=0, v0=32],pos);
(%o) 37.5
```

So the answer is 37.5 m. The substitutions for *a*, x_0, and v_0 are not permanently stored. They are used only to evaluate the expression in this particular instance. If we ask *Maxima* to display the value of *pos*, it will just return to the original expression that we assigned to that variable.

```
(%i) pos;    (%o) x0 + t v0 + \frac{a t^2}{2}
```

A different way to handle this situation is to define a *function*. A function always has an argument that is a variable, although there may be other variables in the function that are not part of the argument. For example, we can define a function that gives the position of our object as a function of time.

```
(%i) x(t):=x0+v0*t+(1/2)*a*t^2;
(%o) x(t) := x0 + v0t + \frac{1}{2} a t^2
```

Note the use of the "colon-equals" for defining a function. It is now easy to evaluate this function at a particular time, say $t = 5$ s.

```
(%i) x(5)    (%o) x0 + 5 v0 + \frac{25 a}{2}
```

We can use the subst command as before to substitute specific values for *a*, x_0 and v_0.

```
(%i) subst([x0=0, a=-9.8, v0=32], x(5));
(%o) 37.5
```

If those values for *a*, x_0, and v_0 are only going to be used for a single calculation, then the substitution method shown above is probably the best way to proceed because we don't want those values permanently stored in those variables. However, if we intend to do many calculations that all use the same set of values for *a*, x_0, and v_0, then it may be more convenient to assign values to those variables permanently.

```
(%i) x0:0$ v0:32$ a:-9.8$
```

Now when we evaluate our function $x(t)$, or our variable *pos*, *Maxima* will replace the variables *a*, x_0, and v_0 with their assigned values. For example, the code below shows that the position of our object at $t = 2.6$ s is 50.076 m. Note the use of the quote–quote operator (' ') in the third command to force subst to evaluate the arguments of *pos*.

```
(%i) [x(2.6), subst(t=2.6, pos), subst(t=2.6,"pos)];
(%o) [50.076, x0 + 2.6 v0 + 3.38 a, 50.076]
```

1.5   2D Kinematics: Vectors

To move beyond one-dimensional motion we will need to deal with vectors. In *Maxima*, vectors are represented as *lists*. A list is just an ordered collection of numbers (or variables, etc.). In *Maxima* lists are always enclosed within square brackets. For example, the code below shows how to define the vector $\vec{V}$ with components $V_x = 2$ and $V_y = -4$ and how to multiply this vector by the scalar q.

```
(%i) V: [2, -4]; q*V;      (%o) [2, −4]    (%o) [2 q, −4 q]
```

If we know the magnitude and direction of a vector, we can use *Maxima* to find the vector components. For example, we can find components for a vector $\vec{A}$ with magnitude $|\vec{A}| = 5.3$ and direction angle $\theta_A = 27°$ and a vector $\vec{B}$ with $|\vec{B}| = 8.1$ and $\theta_B = 230°$. (Note that in the code below we define a constant to convert from degrees to radians, because *Maxima* expects arguments of trigonometric functions to be in radians.)[6]

```
(%i) kill(all)$ deg:%pi/180$
     A:5.3*[cos(27*deg), sin(27*deg)];
     B:8.1*[cos(230*deg), sin(230*deg)];
```
$$(\%o)\ \left[5.3\cos\left(\tfrac{3\pi}{20}\right), 5.3\sin\left(\tfrac{3\pi}{20}\right)\right]\ \left[8.1\cos\left(\tfrac{23\pi}{18}\right), 8.1\sin\left(\tfrac{23\pi}{18}\right)\right]$$

Once our vectors are defined it is easy to carry out vector calculations such as adding the two vectors, or taking the scalar (dot) product of the two vectors. Note that *Maxima* always gives an exact answer when possible, so if we want an (approximate) decimal answer, we must use the `float` command to convert the exact value into a floating-point (decimal) value.

```
(%i) A+B;
```
$$(\%o)\ \left[8.1\cos\left(\tfrac{23\pi}{18}\right) + 5.3\cos\left(\tfrac{3\pi}{20}\right), 8.1\sin\left(\tfrac{23\pi}{18}\right) + 5.3\sin\left(\tfrac{3\pi}{20}\right)\right]$$
```
(%i) float(A+B);
```
$$(\%o)\ [-0.48425,\ -3.7988]$$
```
(%i) float(A.B);     (%o) −39.517
```

We can also combine vectors to create new vector expressions. For example, the code below illustrates one way of defining a vector that gives the position of a projectile as a function of time. First we define vectors for the initial velocity, initial position, and acceleration. Then we can use our knowledge of motion with constant acceleration to combine these vectors into an expression that gives the position vector at time t.

---

[6]We often begin a new example with the `kill(all)` command. This command breaks all connections–assignments of values to name and function calls in particular. Doing this keeps forgotten assignments from contaminating the current analysis. In some settings `kill(all)` appears to affect *Maxima*'s behavior. If you encounter such a situation, replace `kill(all)` with `kill(values, functions, arrays)`. Actually, you can be quite specific with `kill`. For example, `kill(y)` would unbind the name y from whatever expression it is currently assigned, but would leave other values, functions, etc. intact.

```
(%i) v0:v0*[cos(θ),sin(θ)];
(%o) [cos(θ) v0, sin(θ) v0]
(%i) x0:[x0,y0];        (%o) [x0,y0]
(%i) a:[0,-g];          (%o) [0,-g]
(%i) x:x0+v0*t+(1/2)*a*t^2;
```
$$(\%o) \quad [x0 + t\cos(\theta)\, v0, y0 + t\sin(\theta)\, v0 - \tfrac{g t^2}{2}]$$

To select a particular component of a vector we can use square brackets and an argument that specifies the position of the component we wish to select.

```
(%i) x[2];
```
$$(\%o) \quad y0 + t\sin(\theta)\, v0 - \tfrac{g t^2}{2}$$

1.6 Projectile Motion: Solving Equations

To investigate the motion of a projectile in detail it will be convenient to define two separate functions that give the *x*- and *y*-coordinates of the projectile as a function of time.

```
(%i) kill(all)$ x(t):=x0+t*cos(theta)*v0;
     y(t) := y0 + t*sin(theta)*(v0)-g*t^2/2;
```
$$(\%o) \quad x(t) := x0 + t\cos(\theta)\, v0 \qquad y(t) := y0 + t\sin(\theta)\, v0 + \frac{(-g)\, t^2}{2}$$

We can now use one of *Maxima*'s equation solvers to solve for the time when the projectile hits the ground. To obtain an analytical solution for an algebraic equation we can use the `solve` command.

```
(%i) solnt:solve(y(t)=0,t);
```
$$(\%o) \quad [t = -\frac{\sqrt{2g y0 + \sin(\theta)^2\, v0^2} - \sin(\theta)\, v0}{g}, \qquad t = \frac{\sqrt{2g y0 + \sin(\theta)^2\, v0^2} + \sin(\theta)\, v0}{g}]$$

Two solutions are given. *Maxima* provides the mathematical solution to the equation, not the solution to the physical problem. The equation is quadratic in time, so there are two solutions for *t*. However, only the $t > 0$ solution is physically sensible. This problem provides a simple illustration for a general rule about using *Maxima* to solve physics problems: *Maxima* provides a solution to a mathematical problem, but the user must interpret that solution to determine how it applies to the physics problem being solved. The output list is assigned the name `solnt`, which we use below.

Once we have selected the correct (positive) solution, we can assign this value to a name in *Maxima*. That way we can use the solution whenever we need to. Note that we can copy the expression from the solution output above and paste it into the expression for defining our new constant. Alternatively, we can make use of the fact that the expression is stored in *Maxima*'s memory. The input below selects the right-hand side of the second item of `solnt`. It assigns the name `expr_ti` to this solution.

```
(%i) expr_ti:rhs(solnt[2]);
```
$$(\%o) \quad \frac{\sqrt{2g y0 + \sin(\theta)^2\, v0^2} + \sin(\theta)\, v0}{g}$$

Now we can evaluate $x(t)$ at the time of impact to determine the range of the projectile.

(%i) x(expr_ti);   (%o) $\dfrac{\cos(\theta)\,v0\left(\sqrt{2\,g\,y0+\sin(\theta)^2\,v0^2}+\sin(\theta)\,v0\right)}{g} + x0$

Now suppose we want to know the time of impact and the range of the projectile for a particular set of initial conditions. We want to re-evaluate the results we found above, but this time using specific values for the constants (x_0, v_0, etc.). If we plan to explore many different launch conditions, we might wish to assign values to initial conditions in a temporary way, but we may want to assign permanent values to other constants like g. The code below shows one way of calculating the time of impact for a projectile launched from $x_0 = 0$ and $y_0 = 12$ m, with initial speed $v_0 = 22$ m/s and launch angle $\theta = 55°$. Note the use of the quote–quote operator to force *Maxima* to evaluate the expression.

(%i) deg:%pi/180$ g:9.8$ calc_ti: subst([x0=0,
       y0=12, v0=22, theta=55*deg], "expr_ti);
(%o) $.10204\left(\sqrt{484\sin\left(\frac{11\pi}{36}\right)^2 + 235.2} + 22\sin\left(\frac{11\pi}{36}\right)\right)$
(%i) float(%);   (%o) 4.2536

In the last step above we have used the special symbol "%", which always refers to the output of the last command (chronologically, not spatially) that was entered. We find that the projectile is in the air for about 4.25 s.

Finally, we can determine the range of our projectile.

(%i) subst([x0=0,y0=12,v0=22,theta=55*deg],x(calc_ti));
(%o) $2.2449\cos\left(\frac{11\pi}{36}\right)\left(\sqrt{484\sin\left(\frac{11\pi}{36}\right)^2 + 235.2} + 22\sin\left(\frac{11\pi}{36}\right)\right)$
(%i) float(%);   (%o) 53.674

The projectile travels almost 54 m before landing.

1.7 Position, Velocity, and Acceleration: Calculus

Maxima can perform symbolic calculus operations. For example, we can define a function in *Maxima* and then evaluate derivatives and integrals of that function. As an example, consider the function that gives the y-coordinate of our projectile.

(%i) kill(all)$ y(t):= y0+vy0*t-(1/2)*g*t^2;
(%o) $y(t) := y0 + vy0\,t + \left(-\frac{1}{2}\right)g\,t^2$

We can evaluate the y-component of the projectile's velocity by finding the derivative of $y(t)$ with respect to time using *Maxima*'s diff command. The arguments of the diff command include a function and the variable with respect to which we are differentiating.

(%i) diff(y(t),t);   (%o) $vy0 - g\,t$
(%i) vy(t):= "(diff(y(t),t));   (%o) $vy(t) := vy0 - g\,t$

In the second line of code above we define a new function, $v_y(t)$, which gives the y-component of the projectile's velocity. We define the function as the derivative of $y(t)$, but we must use the quote–quote operator and enclose the `diff` command in parentheses in order to force *Maxima* to evaluate the derivative and then use the result to define $v_y(t)$. Without the quotes and parentheses *Maxima* would just define $v_y(t)$ as an abstract derivative of $y(t)$ without actually evaluating that derivative. We can now use this new function to solve for the time when the projectile reaches its peak height (when $v_y = 0$), and then evaluate $y(t)$ at this time to determine the peak height for the projectile.

(%i) `solve(vy(t)=0,t);`   (%o) $[t = \frac{vy0}{g}]$

(%i) `y(vy0/g);`   (%o) $y0 + \frac{vy0^2}{2g}$

If we wish to evaluate the y-component of the projectile's acceleration, we can do so in two ways: we can take the second derivative of $y(t)$ with respect to t or we can take the first derivative of $v_y(t)$ with respect to t. To take higher order derivatives we just insert an additional argument that specifies the order.

(%i) `diff(y(t),t,2);`   (%o) $-g$

(%i) `diff(vy(t),t);`   (%o) $-g$

Maxima can integrate as well. Integrating the acceleration of our projectile should tell us the change in the projectile's velocity. The y-component of acceleration is $-g$ so we can find the change in the y-component of velocity by integrating $-g$ with respect to time. For example, we could determine the change in the projectile's y-component of velocity from $t = 2$ s to $t = 5$ s by evaluating the corresponding definite integral using *Maxima*'s `integrate` command. The arguments of `integrate` are a function, the variable of integration, and the lower and upper limits of integration, respectively.

(%i) `integrate(-g,t,2,5);`   (%o) $-3g$

In looking at the result above it is important to remember that the "3" has units of seconds, so $-3g$ has units of velocity (m/s if g is measured in m/s^2). If we want a general expression for the y-component of velocity, we could evaluate an indefinite integral of the acceleration. To evaluate an indefinite integral we just leave out the limits of integration in the arguments of `integrate`.

(%i) `integrate(-g,t);`   (%o) $-gt$

Here we must be careful because *Maxima* gives the result without a constant of integration. We must remember to add the constant of integration ourselves, using our knowledge of the initial velocity of the projectile. Likewise, we can perform two integrations (by nesting one `integrate` command within another) to determine the y-coordinate of the projectile as a function of time from its acceleration. The code below shows how this can be done both with and without attention paid to the constants of integration. Only when the appropriate constants of integration are added do we get the correct result.

```
(%i) integrate(integrate(-g,t),t);      (%o) −gt²/2
(%i) integrate(integrate(-g,t)+vy0,t)+x0;
```
$$(\%o)\ x0 + t\,vy0 - \frac{g t^2}{2}$$

1.8 Newton's Second Law: Solving ODEs

So far we have looked at projectile motion from a kinematic perspective. We know that the projectile will experience a constant acceleration of magnitude g directed downward. By *why* is this so? The fundamental principles that govern the motion of objects in Newtonian physics are known as Newton's Laws of Motion. The first law states that objects will maintain a constant velocity unless they are subject to a nonzero net force. The third law states that forces always come in interaction pairs in which two objects exert forces of equal magnitude and opposite direction on one another. But if an object is subject to a nonzero net force, how will that object move? This question is answered by Newton's Second Law.

Newton's Second Law is often written in the form of an ODE. For the moment let us consider motion in only one dimension. The position of an object can be thought of as a function of time, $x(t)$. If an object is subject to a net force F_{net}, with the sign of F_{net} indicating the direction of the force (in the $+x$ direction if positive, etc.), then Newton's Second Law states that

$$m\frac{d^2x}{dt^2} = F_{net}. \tag{1.2}$$

Note that this equation involves the derivative of x with respect to t. Because x is a function of t only, this derivative is an ordinary derivative rather than a partial derivative. That makes Newton's Second Law an ODE: it is an equation that involves ordinary derivatives of the function $x(t)$. If we know the net force, then we can attempt to solve this ODE. Not all cases will have an analytical solution, but some do and *Maxima* can help us to find solutions in some of these cases. For example, if we are considering projectile motion near the surface of the Earth, then the force on the projectile has magnitude mg (where m is the mass of the projectile) and it is directed downward (which we will take to be the $-y$ direction). If we consider only the y-component of the projectile's motion, then we can represent Newton's Second Law in *Maxima* as shown below. Note that the single quote instructs *Maxima* to recognize the derivative but not to evaluate it.[7]

```
(%i) kill(all)$ N2: m*'diff(y(t),t,2) = -m*g;
```
$$(\%o)\ m\left(\frac{d^2}{dt^2}y(t)\right) = -g\,m$$

---

[7]Thus the single-quote operation is, in a sense, the opposite of the quote–quote operation, which forces evaluation.

Maxima has several tools for solving ODEs. In some cases we must solve the ODE numerically, but in our projectile motion case we can solve it analytically. The easiest way to solve the ODE is with *Maxima*'s `desolve` command. The code below illustrates the use of this command. Note that the arguments of `desolve` are the ODE and the function for which we want a solution. Also, observe that `y(t)` must be entered, not just `y`.

```
(%i) sol1: desolve(N2,y(t));
```
$$(\%o) \; y(t) = t \left(\frac{d}{dt} y(t) \Big|_{t=0} \right) - \frac{gt^2}{2} + y(0)$$

The solution is given as a function of t, but also in terms of the initial values of y and dy/dt. We can use the `atvalue` command to define different, or more convenient, constants to be used in the solution. In an `atvalue` command we specify a quantity (such as y or dy/dt), a specified time, and the value of that quantity at that time. The example below shows how to set the initial y-coordinate to y_0 and the initial y-velocity to v_{y0}. Then the `desolve` command generates a solution using these new constants.

```
(%i) atvalue(diff(y(t),t),t=0,vy0)$
     atvalue(y(t),t=0,y0)$      sol1:desolve(N2,y(t));
```
$$(\%o) \; y(t) = y0 + t\,vy0 - \frac{gt^2}{2}$$

1.9 Range of a Projectile: Root Finding

Above we found an expression for the x-coordinate of our projectile at the time of impact ($t = t_i$ when $y = 0$). We can use this result to define an expression for the range of the projectile as a function of the launch angle θ: $r(\theta) = x(t_i) - x_0$, or

$$r(\theta) = \frac{v_0^2}{g} \cos \theta \left(\sqrt{\frac{2gy_0}{v_0^2} + \sin^2 \theta} + \sin \theta \right). \tag{1.3}$$

We can rewrite this function as $r(\theta) = 2h_{\max} h(\theta, k)$, where $h_{\max} = v_0^2/(2g)$ is how high the projectile would rise if it was fired straight up, $k = y_0/h_{\max}$ and $h(\theta, k)$ is defined in the code below.

```
(%i) kill(all)$ deg:%pi/180$
     h(theta,k):=(cos(theta)*(sqrt(k+sin(theta)^2)+sin(theta)));
```
$$(\%o) \; h(\theta, k) := \cos(\theta) \left(\sqrt{k + \sin(\theta)^2} + \sin(\theta) \right)$$

The function $h(\theta, k)$ is a dimensionless quantity that depends on the dimensionless parameter k. It is often useful to use dimensionless quantities in computational work (and particularly numerical work) because it is unnecessary to keep track of the units for these quantities, since they have none. Note that to find the angle that maximizes the range of our projectile we only need to maximize $h(\theta)$, with the value of k set by initial conditions. We can determine the approximate value of θ

that maximizes $h(\theta)$ by plotting $h(\theta)$. To create a plot we can use *Maxima*'s `draw` package. First, we load the `draw` package.

```
(%i) load(draw);
(%o) /usr/share/maxima/5.29.1/share/draw/draw.lisp"
```

To generate a 2d plot within our *wxMaxima* notebook we can use `wxdraw2d`. We wish to plot $h(\theta, k)$ as an explicit function of θ for a given value of k, so we use the `explicit` command within `wxdraw2d`. The arguments of the explicit command are the function to be plotted, the independent variable, and the minimum and maximum values of the independent variable to be used. The code below produces a plot of $h(\theta, k)$ for $k = 2$ (recall that k is a dimensionless parameter and h is a dimensionless function) for launch angles ranging from $0°$ to $90°$. Note the optional arguments within `wxdraw2d` that specify labels for the axes. These can be placed anywhere within the command. The output from this command is shown in Fig. 1.2.

```
(%i) wxdraw2d(explicit(h(theta*deg,2),theta,0,90),
        xlabel="{/Symbol q} (deg)", ylabel="h")$
```

Inspection of the plot in Fig. 1.2 shows that for $k = 2$ the launch angle that maximizes the range is close to $30°$. However, we can determine a precise value by solving for the value of θ for which the derivative of h with respect to θ is zero. First we define a new function dh, which is the derivative of $h(\theta, k)$ with respect to θ.

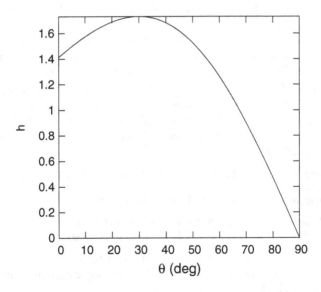

Fig. 1.2 Plot of the dimensionless function $h(\theta, k)$, for $k = 2$. The value of θ that maximizes this function will also maximize the range of the projectile

(%i) **dh(theta,k):="(diff(h(theta,k),theta));**

(%o) $\mathbf{dh}\,(\theta,k) := \cos{(\theta)}\left(\dfrac{\cos(\theta)\sin(\theta)}{\sqrt{\sin(\theta)^2+k}} + \cos{(\theta)}\right) -$

$$\sin{(\theta)}\left(\sqrt{\sin{(\theta)}^2 + k} + \sin{(\theta)}\right)$$

To find the value of θ that maximizes the range we want to solve the equation $dh = 0$ for a given value of k. Unfortunately this equation is transcendental and therefore no analytical solution is possible. We must solve the equation numerically. For this purpose we can use *Maxima*'s find_root command. The arguments of find_root are a function whose roots are to be found, the independent variable, and maximum and minimum values of the variable to be considered. Ideally we want to specify the maximum and minimum values such that one, and only one, root of the function lies between them. For example, to find the value of θ such that $dh = 0$ for $k = 2$ we can search for the root between 20° and 40°.

(%i) **find_root(dh(theta*deg,2),theta,0,90);**   (%o) **30.0**

For $k = 2$ the angle that maximizes the range is $\theta = 30°$. For $k = 5$ we get a different optimal angle.

(%i) **find_root(dh(theta*deg,5),theta,0,90);**
(%o) **22.208**

A larger value of k results in a smaller optimal launch angle. How do we interpret this physically? Recall that $k = y_0/h_{max}$. So a larger value of k means the projectile starts from a launch point that is higher, relative to how far up the projectile would travel if fired straight up. When the projectile is launched from a higher point it makes sense that a smaller (more horizontal) launch angle will maximize the range.

To get a better idea of how the optimal angle changes as k changes we define a new function, $\theta_{max}(k)$, that uses find_root to calculate the optimal angle given a value for k. Once this function is defined we can plot the function and examine the behavior of θ_{max} as k changes. The resulting plot is shown in Fig. 1.3. (Note that we begin the plot at $\theta = 0.01°$ rather than at 0° in order to avoid problems with the find_root command.)

(%i) **theta_max(k):="find_root(dh(theta*deg,k),**
        theta,0,90);
(%o)   **theta_max (k) := find_root (dh (θ deg, k), θ, 0, 90)**

(%i) **wxdraw2d(explicit(theta_max(k),k,0.01,50),**
        xlabel="k",ylabel="{/Symbol q}_max(deg)");

Figure 1.3 makes it clear that the optimal angle for $k = 0$ (launch from the ground) is 45°, but as k increases the optimal launch angle decreases, approaching horizontal ($\theta = 0$) in the limit $k \to \infty$.

Now suppose we want to know the maximum range for a projectile fired at a launch speed of 22 m/s from a height of 15 m above ground level. We can calculate the corresponding value of k (using $g = 9.8$ m/s²), and then find θ_{max} for this value of k. Once we have found the optimal angle we can calculate the maximum range by

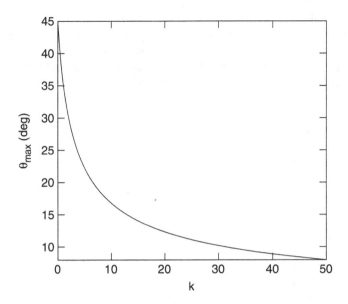

Fig. 1.3 Optimal launch angle versus $k = y_0/h_{max}$

evaluating $2h_{max}h(\theta_{max}) = (v_0^2/g)h(\theta_{max})$. The code below shows that the optimal launch angle for this projectile is approximately $38.3°$, and the maximum range for the projectile is about 62.6 m.

```
(%i) y0:15$ g:9.8$ v0:22$ k:2*g*y0/v0^2$
     ang:theta_max(k);        (%o) 38.264
```

```
(%i)  deg:%pi/180$ (v0^2/g)*h(ang*deg, k);
```
$$(\%o) \quad 49.388\cos(0.21258\,\pi)\left(\sqrt{\sin(0.21258\,\pi)^2 + 0.60744} + \sin(0.21258\,\pi)\right)$$

```
(%i) float((v0^2/g)*h(ang*deg, k));
(%o) 62.616
```

1.10 Visualizing Motion in *Maxima*

In studying the motion of objects we often want a visual picture of how the object moves. We can use *Maxima* to construct plots of motion in a variety of ways. For example, in the case of our projectile we could plot x versus t, y versus t, or y versus x. Each of these plots will help us to visualize different aspects of the projectile's motion. To construct these plots we first define functions for $x(t)$ and $y(t)$.

```
(%i) x(t):= " (x0+t*cos(theta)*v0);
     y(t) := " (y0 + t*sin(theta)*(v0)-g*t^2/2);
```
$$(\%o) \quad x(t) := x0 + t\cos(\theta)\,v0 \quad (\%o2) \quad y(t) := y0 + t\sin(\theta)\,v0 - \frac{gt^2}{2}$$

Next we define our initial conditions. For a launch height of $y_0 = 15$ m and launch speed of $v_0 = 22$ m/s we found that the optimal launch angle was $\theta_{max} = 38.264°$. Using these initial conditions (along with $x_0 = 0$) we construct plots of x versus t (Fig. 1.4), y versus t (Fig. 1.5), and y versus x (Fig. 1.6) below. To keep the output on a single line `fpprintprec` is set to 5.

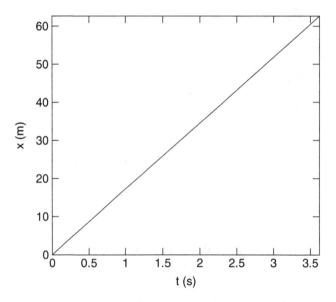

Fig. 1.4 Plot of x versus t

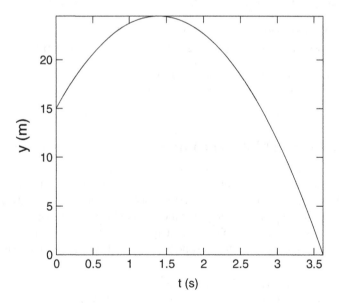

Fig. 1.5 Plot of y versus t

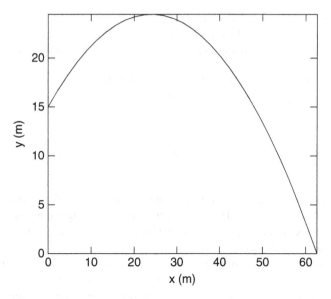

Fig. 1.6 Plot of *y* versus *x*

```
(%i)  x0:0$ y0:15$ v0:22$ g:9.8$ deg:%pi/180$
theta:38.26419432241701*deg$
ti:(sqrt(2*g*y0+sin(theta)^2*v0^2)+sin(theta)*v0)/g; float(%);
```

$$(\%o) \quad .10204\left(\sqrt{484\sin(.21258\,\pi)^2 + 294.0} + 22\sin(.21258\,\pi)\right)$$

```
(%o)    3.625
(%i)  wxdraw2d(explicit(x(t),t,0,ti),xlabel="t (s)",
        ylabel="x (m)")$
(%i)  wxdraw2d(explicit(y(t),t,0,ti),xlabel="t (s)",
        ylabel="y (m)")$

(%i)  wxdraw2d(parametric(x(t),y(t),t,0,ti),
        xlabel="x (m)", ylabel="y (m)");
```

1.11   Exercises

1. Consider the fall of an object from the height of the International Space Station.
   The ISS orbits 370 km above Earth's surface. The Earth is roughly spherical
   with a radius of 6371 km. Assume that the object falls from rest straight down

onto Earth's surface.[8] When objects fall near Earth's surface (in the absence of any resistance), they fall with a constant acceleration of $g = 9.79$ m/s^2. Use this constant acceleration model to examine the motion of the falling object.

(a) The object is moving in just one dimension, along a straight line toward the center of Earth. Let its initial height above ground be h. It falls with a constant acceleration g, so if $y(t)$ represents the height as a function of time we have

$$\frac{d^2}{dt^2} y(t) = -g,$$

where we are defining our y-coordinate so that the positive y-direction is up. Use Maxima's `desolve` to solve this differential equation. Use `atvalue` to set the initial height to h and the initial velocity to zero. Note: the result should look familiar.

(b) Define a new function $y1(t)$ using the solution you just obtained. Use `solve` to find the time of fall before the object hits Earth. Note that you get more than one solution. Which solution is the physically sensible one?

(c) Use `diff` to define a new function $v1(t)$ that gives the object's velocity as a function of time.

(d) Assign appropriate values to the constants g and h. Be careful with units! Compute a numerical value for the time of fall. Determine the speed of the object when it hits the ground.

(e) Plot the height of the object as a function of time. Give appropriate labels to your axes (with units). Use a sensible domain of time values for your plot.

2. Again, consider the fall of an object from the height of the ISS as discussed in the previous question. This time, assume that the force on the object is determined by Newton's Universal Law of Gravitation. So the object experiences a downward force of

$$F = \frac{GMm}{r^2},$$

where $G = 6.674 \times 10^{-11}$ N(m/kg)^2 is Newton's constant, $M = 5.972 \times 10^{24}$ kg is the mass of Earth, m is the mass of the object, and r is the distance of the object from Earth's center. Since we usually measure height from Earth's surface, we will define a new variable y such that $r = R + y$, where R is the radius of Earth.

---

[8]This is not the same as dropping an object from the ISS. An object dropped from the ISS would be orbiting the Earth at the same speed as the ISS, and so it would simply continue to orbit rather than fall to Earth's surface.

(a) Define a function $a(y)$ that gives the acceleration of the object as a function of y (with constants G, M, and R).

(b) It turns out to be pretty complicated to use this expression for the acceleration. However, you can use an approximate expression and still obtain results that are more accurate than assuming constant acceleration. Use the taylor command to find the Taylor series for $a(y)$ about $y = 0$ to fifth order. The syntax for the taylor command is: taylor(f(x),x,a,n), which returns the nth order Taylor series expansion of $f(x)$ about $x = a$. Note that there is a constant term in the Taylor series result. How does this constant relate to the model we considered earlier?

(c) Use desolve to solve the differential equation that we get by keeping only the *two lowest power* terms (the constant term and the y term) in this Taylor series, using initial height $y_0 = h$ and initial velocity zero. Define a new function $y2(t)$ using this solution.

(d) Use diff to define a function $v2(t)$ that gives the velocity as a function of time.

(e) Assign appropriate values to the constants G, M, R, and h. Make a plot of $y2(t)$. Use a sensible domain of time values. Label your axes (with units). Try to estimate when the object hits the Earth.

(f) Use find_root to determine the precise time at which the object hits Earth.

(g) Determine how fast the object is moving when it hits. Compare the results from this model to those from the previous model. Are they very different? Which model do you think is better, and why?

3. Consider an object of mass m moving through a resistive medium in one dimension. The object is subject to a resistive force $F = -bv$, where b is a constant. The initial speed of the object is v_0.

(a) Use desolve to find $x(t)$ for this object. Then use diff to find $v(t)$.

(b) Use the limit command to determine the maximum distance the object can travel. The syntax for the limit command is limit(f(x),x,a) to compute the limit of $f(x)$ as $x \to a$. In *Maxima* the symbol inf is used to represent positive infinity (and minf is used for negative infinity).

(c) Determine the time (in terms of b and m) at which the object has traveled exactly half of its maximum distance. How fast is the object moving (in terms of v_0) at that time?

(d) Construct plots of x (in units of mv_0/b) and v (in units of v_0) as a function of t (in units of m/b).

Chapter 2
Newtonian Mechanics

This chapter addresses Newtonian mechanics. We begin by examining an object that is in *static equilibrium*, where the body, though subject to a set of forces, does not move. We then investigate the motion of an object subject to different types of forces: constant forces, air resistance, and electromagnetic forces.

2.1 Statics

Consider an object that is subject to various forces and torques, but the object does not move in any way. A system of this type is said to be in *static equilibrium*. The branch of mechanics that deals with such systems is known as *statics*. Let's examine a typical problem from statics, using *Maxima* to help us solve the resulting equations.

Consider the situation shown in Fig. 2.1. This diagram shows a 32-kg sign that hangs by a wire from the end of a 20-kg rigid horizontal rod. The other end of the rod is attached to a vertical wall. A second wire connects the end of the rod where the sign is attached to a point higher on the wall. This wire makes an angle $\phi = 35°$ with the rod. Assume that both wires are taut and treat them as massless. What force is exerted on the rod by the wall? What is the tension in the wire that runs from the rod to the wall?

To analyze this situation we take advantage of the fact that the rod is not moving and, therefore, is not accelerating (linearly or rotationally). Newton's second law, and its rotational equivalent, tell us that the net force and net torque on the rod must both be zero. We know that the vertical wire exerts a downward force on the rod that is equal in magnitude to the weight of the sign (because the other end of that wire must be pulling up on the sign with a force equal to the sign's weight). Of course, the rod's own weight also pulls it downward. Denote the tension in the angled wire T and the x- and y-components of the force on the rod by the wall F_{wx} and F_{wy}.

© Todd Keene Timberlake & J. Wilson Mixon, Jr. 2016

T.K. Timberlake, J.W. Mixon, *Classical Mechanics with Maxima*, Undergraduate Lecture Notes in Physics, DOI 10.1007/978-1-4939-3207-8_2

Fig. 2.1 A statics problem: a
sign hangs from a rigid
horizontal rod which has one
end attached to a wall and the
other end connected to the
wall by a wire

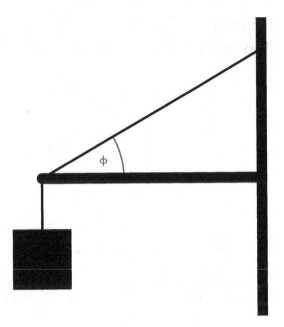

The net force on the rod is zero, so

$$\sum F_x = T\cos(35°) + F_{wx} = 0 \tag{2.1}$$

$$\sum F_y = T\sin(35°) - (32\,\text{kg})(9.8\,\text{m/s}^2) + F_{wy} - (20\,\text{kg})(9.8\,\text{m/s}^2) = 0. \tag{2.2}$$

Also, the sum of the torques about any point on the rod must be zero. For
convenience we choose our origin at the point where the rod touches the wall, as
this will eliminate both F_{wx} and F_{wy} from our torque equation (since they act at
the origin and therefore do not exert any torque about that point). We can treat the
weight of the rod as though it acts at the center of the rod, 1 m from the origin.
The weight of the sign acts at the far end of the rod, 2 m from our chosen origin.
Recalling that counterclockwise torques are considered positive, we have

$$\sum \tau = -T\sin(35°)(2\,\text{m}) + (32\,\text{kg})(9.8\,\text{m/s}^2)(2\,\text{m}) + (20\,\text{kg})(9.8\,\text{m/s}^2)(1\,\text{m}) = 0. \tag{2.3}$$

Solving this system of three equations is not hard, but it is a bit tedious. *Maxima*
can help. First we enter the equations, assigning each equation to a variable.

```
(%i) deg: %pi/180$ g:9.8$ phi:35*deg$
eq1: T*cos(phi)+Fwx=0$
eq2: T*sin(phi)-32*g+Fwy-20*g=0$
eq3: 32*g*2+20*g*1-T*sin(phi)*2=0$
```

Now we use `solve` to solve the system of equations. *Maxima* always tries to give us an exact answer, so whenever possible it will convert decimal values into fractions and it will leave trig functions unevaluated rather than giving a decimal representation. In most cases we desire a decimal value, so we use `float` to have *Maxima* convert the result.

(%i) `sol:solve([eq1,eq2,eq3],[Fwx,Fwy,T]);`

(%o) $\left[\left[\mathrm{Fwx} = -\frac{2058\cos\left(\frac{7\pi}{36}\right)}{5\sin\left(\frac{7\pi}{36}\right)}, \mathrm{Fwy} = 98, \mathrm{T} = \frac{2058}{5\sin\left(\frac{7\pi}{36}\right)}\right]\right]$

(%i) `float(%);`   (%o) $[[\mathrm{Fwx} = -587.83, \mathrm{Fwy} = 98.0, \mathrm{T} = 717.6]]$

The tension in the angled wire is 717.6 N. The force on the rod by the wall has an x-component of $-587.8\,N$ and a y-component of 98 N. To test that this purported solution does solve our system of equations, we instruct *Maxima* to substitute our solution back into each equation. For example, the code below verifies our solution for the first equation.[1]

(%i) `subst(sol[1],eq1);`   (%o) $0 = 0$

In order to set up our torque equation, we had to choose an origin about which the torques would be calculated. However, the solution to our problem should not depend on the choice of origin, because the net torque should be zero about *any* point on the rod. To verify this we can replace our torque equation from above with a new torque equation that uses the left end of the rod as the origin. This eliminates the variable T from the equation (since the tension acts at that end of the rod and thus produces no torque), but it also eliminates F_{wx} because the horizontal component of any force on the rod must act through our origin and therefore produce no torque. The resulting torque equation is

$$\sum \tau = F_{wy}(2\,\mathrm{m}) - (20\,\mathrm{kg})(9.8\,\mathrm{m/s^2})(1\,\mathrm{m}) = 0. \tag{2.4}$$

We can now use *Maxima* to solve the new system of equations consisting of Eqs. 2.2 and 2.4.

(%i) `eq4: Fwy*2-20*g*1=0$`
       `solve([eq1,eq2,eq4],[Fwx,Fwy,T]);`

(%o) $\left[\left[\mathrm{Fwx} = -\frac{2058\cos\left(\frac{7\pi}{36}\right)}{5\sin\left(\frac{7\pi}{36}\right)}, \mathrm{Fwy} = 98, \mathrm{T} = \frac{2058}{5\sin\left(\frac{7\pi}{36}\right)}\right]\right]$

(%i) `float(%);`   (%o) $[[\mathrm{Fwx} = -587.83, \mathrm{Fwy} = 98.0, \mathrm{T} = 717.6]]$

The solution is identical to the one we obtained earlier, as expected. Now we can put a new twist on our problem. Suppose the second wire had a tensile strength of only 1000 N. In order to leave plenty of room for error (or perching birds!) we decide we do not want to exceed a tension of 500 N on this wire, so the tension of 717.6 N with $\phi = 35°$ is too large. What angle ϕ will give us a tension of 500

---

[1]Observe that `sol` consists of a list embedded in another list. The command uses the notation `sol[1]` to extract the list of solutions that is to be placed into `eq1`.

N in the wire? To solve this problem we introduce two variables T_x and T_y which represent the x- and y-components of the tension in the second wire. We choose our origin at the left end of the rod in order to eliminate T_x, T_y and F_{wx} from our torque equation. The resulting equations for zero net force and torque are entered into *Maxima* and solved as shown below. Note that the final equation (eq8) ensures that the magnitude of the tension will be $500\,N$.

```
(%i) kill(phi)$ g:9.8$              eq5:Tx+Fwx=0$
     eq6:Ty-32*g+Fwy-20*g=0$        eq7:Fwy*2-20*g*1=0$
     eq8:Tx^2+Ty^2=500^2$
     sol2:solve([eq5,eq6,eq7,eq8],[Fwx,Fwy,Tx,Ty]);
```
$(\%o)$ $[Fwx = -\frac{2\sqrt{503659}}{5}, Fwy = 98, Tx = \frac{2\sqrt{13}\sqrt{17}\sqrt{43}\sqrt{53}}{5}, Ty = \frac{2058}{5}]$

$[Fwx = \frac{2\sqrt{503659}}{5}, Fwy = 98, Tx = -\frac{2\sqrt{13}\sqrt{17}\sqrt{43}\sqrt{53}}{5}, Ty = \frac{2058}{5}]$

Note that two different solutions are given, but the difference between the two solutions is just a change in sign of the x-components of the tension and the force from the wall. In this problem it makes no physical sense for the tension to have a negative x-component (the wire can pull, but it cannot push) so we can ignore the second solution. Now we can find the angle for ϕ with simple trigonometry. We know that the tension must point along the wire, so

$$\phi = \tan^{-1}\left(\frac{T_y}{T_x}\right). \tag{2.5}$$

We use *Maxima* to evaluate the inverse tangent function.

```
(%i) subst(sol2[1],atan(Ty/Tx));    (%o) atan(1029/(√13 √17 √43 √53))
```

```
(%i) float(%*180/%pi);    (%o)  55.406
```

The wire has a tension of $500\,N$ when $\phi \approx 55.4°$. Thus, in order to reduce the tension we needed to increase ϕ. We might wonder what is the minimum possible tension in this wire? To achieve the minimum tension we would need to maximize the angle, so the wire would need to be vertical (or as close to vertical as possible so that it can still connect to the wall). If we assume $\phi = 90°$ we can use *Maxima* to find T_y ($T_x = 0$ since the wire is going straight up). If we calculate torque about the left end of the rod then our equations are

$$\sum F_y = T_y - (32\,\text{kg})(9.8\,\text{m/s}^2) + F_{wy} - (20\,\text{kg})(9.8\,\text{m/s}^2) = 0, \tag{2.6}$$

$$\sum \tau = F_{wy}(2\,\text{m}) - (20\,\text{kg})(9.8\,\text{m/s}^2)(1\,\text{m}) = 0. \tag{2.7}$$

We use *Maxima* to solve this system below.

```
(%i) g:9.8$              eq9:Ty-32*g+Fy-20*g=0$
     eq10: Fy*2-20*g*1=0$
     sol3:solve([eq9,eq10],[Fy,Ty]);
(%o) [[Fy = 98, Ty = 2058/5]]
```

The minimum tension (for a vertical wire) is 2058/5 N, or 411.6 N. Note that this result is just the weight of the sign plus half the weight of the rod. If the wire pulls up vertically on the end of the rod then it must support the weight of the sign and also cancel the torque caused by the weight of the rod. If we use the contact point with the wall as the origin, it is not hard to see that the wire must pull up with half the weight of the rod in order to offset the torque due to the rod's weight.

2.2 Constant Forces: Block on a Wedge

In our statics example, the forces on the rod were constant but they all canceled out resulting in no acceleration. Now we examine a case where the forces are still constant, but they do not necessarily cancel out. Here we must allow for the objects involved to accelerate, and one of our main goals will be to determine the acceleration of the objects.

Consider a rectangular block of mass m placed on a triangular wedge that has a mass M and an incline angle θ. The wedge, in turn, sits upon a horizontal surface. The arrangement is shown in Fig. 2.2. There is no friction between the block and the wedge, or between the wedge and the horizontal surface. The block and the wedge are both subject to constant gravitational forces, and they can also exert constant normal forces on each other. These *normal forces* must be equal in magnitude and opposite in direction according to Newton's Third Law.

We start by defining coordinates. We can ignore the vertical position of the wedge because that will never change. However, we need a coordinate to specify the horizontal position of the wedge. Let X measure the horizontal distance from the origin O to the corner of the wedge as shown in Fig. 2.2. We will require two coordinates, x and y, to represent the position of the bottom edge of the block relative to the origin as shown in Fig. 2.2.

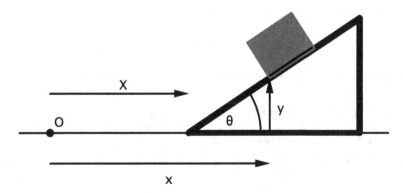

Fig. 2.2 A block sitting on a moveable wedge

The block is subject to two forces: its own weight and a normal force from the wedge. The weight has magnitude mg and is in the negative y-direction. Let N represent the magnitude of the normal force, which must point perpendicular to the surface of the wedge. This force has a negative x-component of magnitude $N \sin \theta$ and a positive y-component of magnitude $N \cos \theta$. If a_x and a_y represent the x- and y-components of the block's acceleration, then by Newton's Second Law we have

$$- N \sin \theta = ma_x,$$

$$N \cos \theta - mg = ma_y. \tag{2.8}$$

The forces on the wedge include both its weight, which is in the negative y-direction, and the normal force from the block which has both x- and y-components (opposite those of the normal force on the block by the wedge). We have already seen that we can ignore the y-motion of the wedge (since there isn't any), so we need only consider the x-components of the forces on the wedge. Newton's Second Law then gives

$$N \sin \theta = MA_X, \tag{2.9}$$

where A_X is the acceleration of the wedge.

Examining Eqs. 2.8 and 2.9 we see that we have three equations, but four unknowns (a_x, a_y, A_x, and N). We need a fourth equation in order to solve the system. The fourth equation comes from the relationship between the coordinates x, y, and X. The motion of the block is *constrained* to be along the surface of the wedge, and we must account for this constraint. From basic trigonometry we see that if the block is on the wedge then we must have

$$(x - X) \tan \theta = y. \tag{2.10}$$

Equation 2.10 is called the equation of constraint. We can differentiate equation 2.10 twice with respect to time to get

$$(a_x - A_X) \tan \theta = a_y. \tag{2.11}$$

Now we can use *Maxima* to solve the system of equations in Eqs. 2.8, 2.9, and 2.11.

```
(%i)  kill(all)$   eq1:-N*sin(theta)=m*ax$
      eq2:N*cos(theta)-m*g=m*ay$   eq3:N*sin(theta)=M*Ax$
      eq4:tan(theta)*(ax-Ax)=ay$
      sol:solve([eq1,eq2,eq3,eq4],[N,ax,ay,Ax]);
(%o)  N = -------- g m M --------
          sin(θ)(tan(θ)M+m tan(θ))+cos(θ)M
      ax = - ----------- g sin(θ) M -----------
             sin(θ)(tan(θ)M+m tan(θ))+cos(θ)M
      ay = - -------- sin(θ)(g tan(θ)M+g m tan(θ)) --------
             sin(θ)(tan(θ)M+m tan(θ))+cos(θ)M
      Ax = ----------- g m sin(θ) -----------
           sin(θ)(tan(θ)M+m tan(θ))+cos(θ)M
```

The result looks quite complicated. However, we can get *Maxima* to simplify the result using `trigsimp`, which will take advantage of some basic trigonometric identities to rewrite the solution in simpler form.

(%i) `nsol:trigsimp(sol);`

(%o) $N = \frac{g\,m\cos(\theta)\,M}{M+m\sin(\theta)^2}$ $\qquad ax = -\frac{g\cos(\theta)\sin(\theta)\,M}{M+m\sin(\theta)^2}$

$ay = -\frac{g\sin(\theta)^2\,M+g\,m\sin(\theta)^2}{M+m\sin(\theta)^2}$ $\qquad Ax = \frac{g\,m\cos(\theta)\sin(\theta)}{M+m\sin(\theta)^2}$

Any time we obtain an analytical solution for a physics problem it is a good idea to test the solution in certain limits where we know, or can easily determine, the answer. For example, if the wedge were flat ($\theta = 0$) then we know that neither the block nor the wedge would accelerate and the normal force would simply equal the weight of the block. We can test this limit in *Maxima* by evaluating our solution when $\theta = 0$.

(%i) `subst( theta=0,nsol);`

(%o) $[[N = g\,m,\ ax = 0,\ ay = 0,\ Ax = 0]]$

The results fit with our expectations. Likewise we could evaluate the solution for $\theta = \pi/2$, in which case the wedge is vertical and the block will be in freefall.

(%i) `subst(theta=%pi/2,nsol);`

(%o) $[[N = 0,\ ax = 0,\ ay = -\frac{g\,M+g\,m}{M+m},\ Ax = 0]]$

It is easy to see that this result simplifies to $a_y = -g$ with all other quantities zero as expected. We can also evaluate our solution in some more interesting limits. For example, if the mass of the wedge is much greater than the mass of the block, what will happen? We can use *Maxima*'s `limit` command to evaluate our solution in the limit $M \rightarrow \infty$. The arguments of `limit` are the expression to be evaluated, the variable whose limit we are taking, and the limiting value of that variable, respectively. Note that positive infinity is represented as `inf` in *Maxima*.

(%i) `limit(nsol,M,inf);`

(%o) $[[N = g\,m\cos(\theta),\ ax = -g\cos(\theta)\sin(\theta),\ ay = -g\sin(\theta)^2,\ Ax = 0]]$

These results require some interpretation, but if we use these values for a_x and a_y to evaluate the magnitude and direction of the block's acceleration we will find that $|\vec{a}| = g\sin\theta$ with the acceleration directed down the slope of the wedge. The wedge itself remains stationary (assuming it started from rest). This is exactly the result for a block sliding down a frictionless inclined plane, as we should expect in this case.

Finally, consider the case in which the block is much more massive than the wedge. We evaluate our solution in the limit $m \rightarrow \infty$.

(%i) `limit(nsol,M,inf);`

(%o) $[[N = \frac{g\cos(\theta)\,M}{\sin(\theta)^2},\ ax = 0,\ ay = -g,\ Ax = \frac{g\cos(\theta)}{\sin(\theta)}]]$

Here we see that the wedge supplies no resistance to the block, so the block is in freefall. However, the block pushes the wedge horizontally, giving it an acceleration $A_X = g\cot\theta$. In effect, the block falls and as it does so it shoots the wedge out sideways.

2.3 Velocity-Dependent Forces: Air Resistance

Introductory physics courses generally ignore the effects of air resistance. This section examines a simple model for air resistance and explores how air resistance can affect the motion of a falling particle or a projectile.

Air resistance is a force that opposes the motion of an object through air. This force exists because the object moving through the air collides with air molecules. These collisions tend to slow the object down, so this force is sometimes referred to as a *drag*. The force of air resistance always points in a direction opposite to the object's velocity. The magnitude of the force depends on the object's speed and its shape.

2.3.1 Models of Air Resistance

One simple, but useful, model for air resistance is represented by the equation

$$\vec{F} = -f(v)\hat{v} \tag{2.12}$$

where v is the speed of the object and $\hat{v}$ is a unit vector in the direction of the object's velocity. The function $f(v)$ is defined by

$$f(v) = bv + cv^2 \tag{2.13}$$

where b and c are constants. For a spherical object, $b = \beta D$ and $c = \gamma D^2$ where D is the diameter of the object. For motion through air at standard temperature and pressure we will use $\beta = 1.6 \times 10^{-4}\,\mathrm{N\,s/m^2}$, and $\gamma = 0.25\,\mathrm{N\,s^2/m^4}$.

The function $f(v)$ has a linear term and a quadratic term. The ratio of these two terms,

$$\frac{f_{quad}}{f_{lin}} = \frac{\gamma D v}{\beta} \approx (1.6 \times 10^3\ \mathrm{s/m^2}) D v, \tag{2.14}$$

reveals that the quadratic term tends to be larger for large diameter objects moving at high velocities, and the linear term dominates for small diameter objects moving at low velocities.

To get some idea of how these terms affect drag at different speeds, consider a 10 cm sphere. Here $D = 0.1$ m, so $b = 1.6 \times 10^{-5}\,\mathrm{N\,s/m}$ and $c = 2.5 \times 10^{-3}\,\mathrm{N\,s^2/m^2}$. We can plot $f(v)$, bv, and cv^2 against v over the range $0 < v < 0.001$ m/s in order to compare the magnitudes of the linear and quadratic terms. Note the use of various options within the wxdraw2d command in the code below. These options set the width of the plotted line, the color of that line, the labels to be used in the key, and the

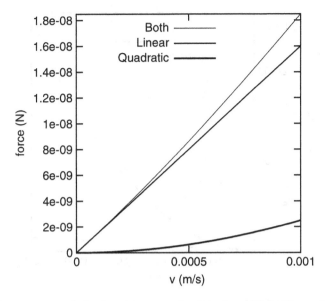

Fig. 2.3 Linear versus quadratic air resistance as a function of speed for a 10 cm sphere

location of the key, as well as the *x*- and *y*-axis labels we have encountered before. For more details about these options consult the *Maxima* manual. The resulting plot is shown in Fig. 2.3.

```
(%i) f(v, b, c) := b*v + c*v^2$ b1:1.6e-5$ c1:2.5e-3$
     wxdraw2d(line_width=1,key="Both",
      explicit(f(v,b1,c1),v,0,0.001),line_width=2,
      key="Linear",explicit(f(v,b1,0),v,0,0.001),
      color=gray50,key="Quadratic",explicit(f(v,0,c1),
      v,0,0.001), xlabel="v (m/s)",ylabel="force (N)",
      user_preamble="set key left")$
```

In the range of speeds shown, the air resistance force is dominated by the linear term with the quadratic term making only a small addition at the high end of the range.

If we modify the code above to generate a plot over the range $0 < v < 0.1$ we get a very different picture. Figure 2.4 shows that at higher speeds (say, $v > 0.05$ m/s) the quadratic term dominates. So for low speeds we can obtain an accurate picture of the motion while ignoring the quadratic term. At high speeds we get accurate results while ignoring the linear term. Below we will examine cases of motion with linear resistance only, and motion with quadratic resistance only. In some situations we must use both the linear and quadratic terms. Examining the effects of the combined linear and quadratic forces is left as an exercise for the reader.

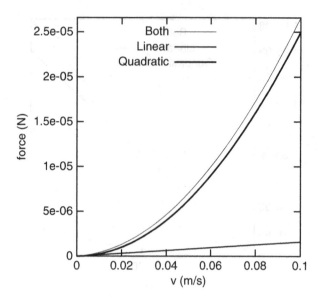

Fig. 2.4 Linear and quadratic resistance, extended range

2.3.2 *Falling with Linear Resistance*

We first consider an object that is falling in Earth's gravitational field and that is subject to linear air resistance. As we have seen, a linear model for air resistance is appropriate for small objects moving at relatively slow speeds. Newton's second law for this object is:

$$m\ddot{y} = -mg - b\dot{y}, \tag{2.15}$$

where $\ddot{y} = d^2y/dt^2$ is the acceleration, $\dot{y} = dy/dt$ is the velocity, and y is the object's distance above the Earth's surface.

We can break this second order differential equation up into two first order ordinary differential equations (ODEs):

$$m\dot{v} = -mg - bv,$$
$$\dot{y} = v. \tag{2.16}$$

We can then use desolve to solve the first ODE.

```
(%i) kill(values, functions, arrays)$
     eq1:m*'diff(v(t),t)=-m*g-b*v(t)$
     sol:desolve(eq1,v(t));
```
$$(\%o) \quad v(t) = \frac{(gm^2 + v(0)\,bm)\,e^{-\frac{bt}{m}}}{bm} - \frac{gm}{b}$$

We assume that the object is dropped from rest, so $v(0) = 0$. We then simplify the solution to the ODE and define a new function for $v(t)$.

```
(%i)   v(t) := (g*m/b) * (%e^(-(b*t)/m)-1);
```
$$(\%o) \quad v(t) := \frac{gm}{b}\left(e^{\frac{-bt}{m}} - 1\right)$$

Note that $v(0) = 0$ as it should. It is easy to show that $\lim_{t\to\infty} v(t) = -gm/b$. This result indicates that our dropped object has a *terminal speed* of gm/b. When the object is released, gravity is pulling it downward and there is no air resistance, so the object will fall. As the object falls, it speeds up. As it speeds up, however, the magnitude of the air resistance force increases. Eventually the object will be falling so fast that the magnitude of the air resistance will equal the object's weight. At this point the two forces will cancel each other and the object will be in equilibrium, so it no longer accelerates. From this point onward the object will fall with a constant speed, which is just the terminal speed we found above. Another way to determine this terminal speed is to set the force of air resistance equal to the object's weight and solve for the speed:

$$b|v| = mg \rightarrow |v| = \frac{mg}{b}. \tag{2.17}$$

The terminal speed obtained in this way is identical to the speed we obtained from the infinite time limit of $v(t)$.

Now we can determine $y(t)$. To obtain $y(t)$ we integrate $v(t)$ with respect to time. Remember, though, that an undetermined constant must be added to this integral to yield the correct solution for $y(t)$. We can determine the value of this constant once we know the initial conditions, but first let's perform the integration. We will also use the expand command to expand the expression by multiplying out all of the terms.

```
(%i) integrate(v(t),t);
```
$$(\%o) \quad \frac{gm\left(-\frac{me^{-\frac{bt}{m}}}{b}-t\right)}{b}$$

```
(%i) expand(%);
```
$$(\%o) \quad -\frac{gm^2 e^{-\frac{bt}{m}}}{b^2} - \frac{gmt}{b}$$

Integration yields $-gm^2/b^2$ at $t = 0$. If we let h be the initial height of the object then $y(0) = h$ and we must add $gm^2/b^2 + h$ to the result of the above integral in order to get the correct expression for $y(t)$. Now we can define the function for $y(t)$.

```
(%i) y(t) := -(g*m^2*%e^(-(b*t)/m))/b^2-
             (g*m*t)/b + h;
```
$$(\%o) \quad y(t) := \frac{-gm^2 e^{\frac{-bt}{m}}}{b^2} - \frac{gmt}{b} + \frac{gm^2}{b^2} + h$$

Consider a specific example. Suppose a rain drop falls from rest from a cloud that is 2 km above Earth's surface. The diameter of the drop is 2 mm and its mass is 4.2×10^{-6} kg. Recall that $b = \beta D$, where D is the diameter of a spherical object and $\beta = 1.6 \times 10^{-4}$ N s/m². We can use the code below to calculate b for the rain drop and then construct a plot of y versus t to determine the approximate time that the

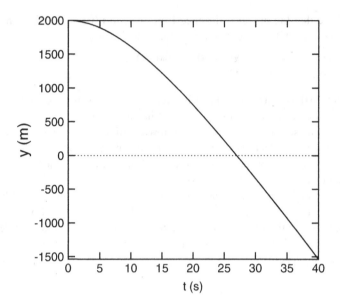

Fig. 2.5 The height of a raindrop (falling with linear resistance) as a function of time

drop will hit the ground. The resulting plot is shown in Fig. 2.5. The `xaxis=true` option produces a dotted line along the *x*-axis in the plot.

```
(%i)  D:2e-3$ m:4.2e-6$ beta:1.6e-4$ b:beta*D$
      g:9.8$ h:2000$
      wxdraw2d(xaxis=true, xlabel="t (s)",
      ylabel = "y (m)", explicit(y(t),t,0,40))$
```

The drop hits the ground ($y = 0$) somewhere between $t = 25$ and $t = 30$ s. We can use `find_root` to determine precisely when the drop hits the ground.

```
(%i)  tend:find_root(y(t),t,25,30);      (%o)   26.996
```

Thus, the drop strikes the ground at $t \approx 27$ s. We can also plot the velocity of the drop as it falls, using the code below. Figure 2.6 shows the resulting plot.

```
(%i) wxdraw2d(ylabel="v (m/s)",
      xlabel = "t (s)", explicit(v(t),t,0,tend))$
```

The effects of air resistance are clear in this case. The rain drop's speed is increasing as it falls, but the speed does not increase linearly. The speed is beginning to level off (approach a maximum negative value) as the drop approaches the ground. We can determine the object's velocity upon impact.

```
(%i)  v(tend);      (%o)  -112.18
```

The rain drop is moving downward at 112.2 m/s when it hits the ground. Let's compare that speed to the terminal speed of the drop. Recall that the terminal speed is gm/b. (Note: using the single quote operator to preclude computation on the

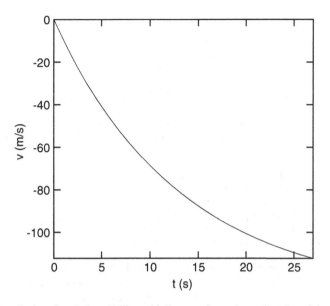

Fig. 2.6 The velocity of a raindrop (falling with linear resistance) as a function of time

left-hand side in the code below is required. The command $g*m/b$ would yield the result, 128.63. Of course, the expression on the left-hand side is not required, but it does provide context.)

(%i)   ' (g*m/b) = g*m/b;   (%o) $\frac{gm}{b} = 128.63$

The terminal speed in this case is 128.6 m/s. In our example above the rain drop reaches an appreciable fraction of its terminal speed before hitting the ground. The only reason it doesn't make it all the way to terminal speed is that it hits the ground before it can do so. If we were to let the drop fall for a longer period of time we would find that it reaches its terminal speed (or very close to it). To verify this fact we can evaluate the rain drop's velocity at $t = 100$ s.

(%i)   v(100);   (%o) -128.56

The speed of the rain drop after 100 s of fall is very nearly equal to the drop's terminal speed.

2.3.3 Projectile Motion with Linear Resistance

We now consider a projectile near Earth's surface and subject to linear air resistance. The equations of motion for projectile are:

$$\dot{x} = v_x,$$

$$\dot{y} = v_y,$$

$$\dot{v}_x = -bv_x,$$

$$\dot{v}_y = -mg - bv_y. \tag{2.18}$$

We apply `desolve` to this system of four ODEs, with `atvalue` used to specify the initial conditions for an object launched from the origin with speed v_0 at a launch angle θ.

```
(%i)  eq1:vx(t)='diff(x(t),t)$
      eq2:vy(t)='diff(y(t),t)$
      eq3:m*'diff(vx(t),t)=-b*vx(t)$
      eq4:m*'diff(vy(t),t)=-m*g-b*vy(t)$
      atvalue(x(t),t=0,0)$
      atvalue(vx(t),t=0,v0*cos(%theta))$
      atvalue(y(t),t=0,0)$
      atvalue(vy(t),t=0,v0*sin(%theta))$         sol2:
      desolve([eq1,eq2,eq3,eq4],[x(t),vx(t),y(t),vy(t)]);
```

$(\%o)$ $x(t) = \frac{\cos(\theta)\,m\,v0}{b} - \frac{\cos(\theta)\,m\,e^{-\frac{bt}{m}}\,v0}{b}$ $vx(t) = \cos(\theta)\,e^{-\frac{bt}{m}}\,v0$

$y(t) = -\frac{e^{-\frac{bt}{m}}\left(\sin(\theta)\,b\,m^2\,v0+g\,m^3\right)}{b^2\,m} + \frac{\sin(\theta)\,b\,m\,v0+g\,m^2}{b^2} - \frac{g\,m\,t}{b}$

$vy(t) = \frac{e^{-\frac{bt}{m}}\left(\sin(\theta)\,b\,m\,v0+g\,m^2\right)}{b\,m} - \frac{g\,m}{b}$

We can copy and paste these solutions into new functions so that they can be manipulated within *Maxima*.

```
(%i)  xs(t):=(m*cos(%theta)*v0)/
      b-(m*%e^(-(b*t)/m)*cos(%theta)*v0)/b$
      vxs(t):=%e^(-(b*t)/m)*cos(%theta)*v0$
      ys(t):=-(%e^(-(b*t)/m)*(b*m^2*sin(
      %theta)*v0+g*m^3))/(b^2*m)+(b*m*sin(%theta)*
      v0+g*m^2)/b^2-(g*m*t)/b$
      vys(t):=(%e^(-(b*t)/m)*(b*m*sin(%theta)*
      v0+g*m^2))/(b*m)-(g*m)/b$
```

Consider the specific case of the motion of the rain drop that we analyzed in the previous example. This time, however, imagine the water drop is fired from a squirt gun at a 45° angle. The strongest water guns can fire water at speeds upward of 15 m/s, so we take 15 m/s as our initial velocity. To begin, we plot y versus t in order to see when the water drop lands.

```
(%i)  D:2e-3$ m:4.2e-6$ beta:1.6e-4$
      b:beta*D$ g:9.8$ v0:15$ deg:%pi/180.0$ %theta:45*deg$
      wxdraw2d(xlabel = "t (s)", ylabel = "y (m)",
      xaxis = true, explicit(ys(t),t,0,3))$
```

Figure 2.7 shows that the plot of `ys(t)` crosses the horizontal axis between $t = 2$ and $t = 2.5$ s. We can use `find_root` to more precisely estimate the landing time.

```
(%i)  tend:find_root(ys(t)=0,t,2,2.5);      (%o)   2.1082
```

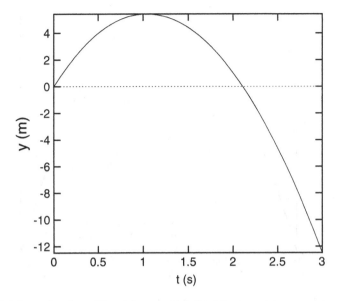

Fig. 2.7 Height as a function of time for a water drop fired from a squirt gun

So this water drop hits the ground at approximately $t = 2.11$ s. With this information, we can plot the trajectory (y versus x) over its entire flight. We also show the trajectory of the water drop without air resistance for comparison. The commands are essentially the same as above, so they are not repeated.

Figure 2.8 shows the effect that air resistance has on the drop's trajectory. Instead of a symmetric parabola, the drop moves along an asymmetric arc and the range of the projectile is reduced. We can find the range for our water drop by evaluating $x(t)$ at the landing time.

```
(%i)   float(xs(tend));       (%o)   20.657
```

This drop lands 20.7 m from where it was launched. Without air resistance the range would be almost 23 m, so air resistance reduces the range by just over 2 m.

2.3.4 Falling with Quadratic Resistance

Now we examine motion with quadratic air resistance. If an object is falling vertically in a medium with quadratic resistance, the equations of motion are

$$m\dot{v} = -mg + cv^2,$$
$$\dot{y} = v. \tag{2.19}$$

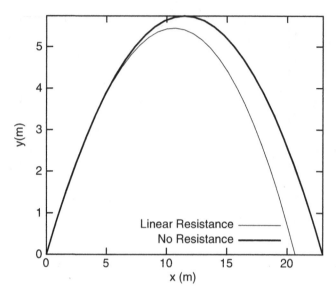

Fig. 2.8 Water drop trajectory with and without air resistance

Note that the resistance term in the first equation is positive because the object is falling (moving in the negative y-direction) and thus the drag force points upward (in the positive y-direction). The first ODE above is nonlinear because of the v^2 term. Since desolve only solves linear ODEs we cannot use desolve to solve this problem. However, *Maxima* has other tools for solving ODEs. For this problem we will use ode2. The ode2 command takes three arguments: the ODE to be solved (which must be first or second order), the dependent variable, and the independent variable, respectively. The code below solves our first ODE using ode2.

```
(%i) kill(c,g,m,v)$ assume(c>0,g>0,m>0)$
     sol:ode2(m*'diff(v,t)=-m*g+c*v^2,v,t);
```

$$(\%o) \quad \frac{\sqrt{m}\log\left(-\frac{\sqrt{c}\sqrt{g}\sqrt{m}-cv}{cv+\sqrt{c}\sqrt{g}\sqrt{m}}\right)}{2\sqrt{c}\sqrt{g}} = t + \%c$$

There are a few things to note about this example. First, the syntax for writing the ODE differs from that used for desolve. Specifically, $v(t)$ is just written as v and the diff command is preceded by an apostrophe to suppress evaluation of the derivative (because otherwise dv/dt would evaluate to zero since *Maxima* will treat v as a constant with respect to t). Note also that we use the assume command to indicate that all of the constants are positive. If this command is left out then *Maxima* will inquire about certain quantities being positive or negative before determining the solution. Finally, the solution includes an undetermined constant denoted by \%c.

We can use another command, ic1, to specify the initial conditions for our solution.

(%i)   icl(sol,t=0,v=0);

(%o) $\dfrac{\sqrt{m}\log\left(-\frac{\sqrt{c}\sqrt{g}\sqrt{m}-cv}{cv+\sqrt{c}\sqrt{g}\sqrt{m}}\right)}{2\sqrt{c}\sqrt{g}} = \dfrac{2\sqrt{c}\sqrt{g}t+\log(-1)\sqrt{m}}{2\sqrt{c}\sqrt{g}}$

The equation above is not yet solved for v as a function of t. We can copy and paste the equation into `solve` and let *Maxima* do the algebra.

```
(%i) solve((m*log(-(sqrt(c*g*m)-c*v)/
     (c*v+sqrt(c*g*m))))/(2*sqrt(c*g*m))=
     (2*sqrt(c*g*m)*t+log(-1)*m)/(2*sqrt(c*g*m)),v);
```

(%o) $\left[\left[v = -\dfrac{\sqrt{g}\sqrt{m}\left(e^{\frac{2\sqrt{c}\sqrt{g}t}{\sqrt{m}}}-1\right)}{\sqrt{c}\left(e^{\frac{2\sqrt{c}\sqrt{g}t}{\sqrt{m}}}+1\right)}\right]\right]$

Recall that

$$\frac{e^x - e^{-x}}{e^x + e^{-x}} = \tanh(x), \tag{2.20}$$

and we can see from the above result that our velocity function can be written as

$$v(t) = -\sqrt{\frac{gm}{c}}\tanh\left(\sqrt{\frac{cg}{m}}t\right). \tag{2.21}$$

We define this $v(t)$ function in *Maxima* and then use the `integrate` command to find $y(t)$. We must be careful to add the appropriate constant to the result of the integral in order to satisfy our initial conditions.

(%i) `v(t):=-sqrt(g*m/c)*tanh(sqrt(c*g/m)*t)$`

`integrate(v(t),t);`   (%o) $-\dfrac{m\log\left(\cosh\left(\frac{\sqrt{c}\sqrt{g}t}{\sqrt{m}}\right)\right)}{c}$

This result evaluates to zero when $t = 0$, so if we want $y(0) = h$, we must add h to the above expression to get the correct $y(t)$. We can then define a new function using this result.

(%i) `y(t):=-(m*log(cosh((sqrt(c)*sqrt(g)*t)/`
     `sqrt(m))))/c+h;`

(%o) $y(t) := \dfrac{-m\log\left(\cosh\left(\frac{\sqrt{c}\sqrt{g}t}{\sqrt{m}}\right)\right)}{c} + h$

We now use our results to examine the motion of the falling raindrop we considered above, but this time with quadratic resistance. Recall that for quadratic resistance $c = \gamma D^2$, where $\gamma = 0.25\,\text{N}\,\text{s}^2/\text{m}^4$. We can define our parameters and plot $y(t)$ to find the time when the drop hits the ground. Figure 2.9 shows the plot.

(%i) `D:2e-3$ m:4.2e-6$ gamma:0.25$ c:gamma*D^2$`
     `g:9.8$ h:2000$`
     `wxdraw2d(xaxis=true, xlabel="t(s)", ylabel= "y(m)",`
     `explicit(y(t),t,0,400))$`

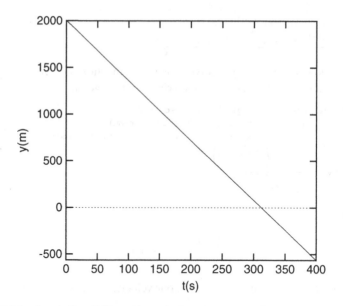

Fig. 2.9 Height of a raindrop (falling with quadratic resistance) as a function of time

Note the linear nature of this graph, suggesting that the drop is falling at a constant velocity. The drop appears to hit the ground between $t = 300$ and $t = 350$ s. We can more precisely determine the landing time using find_root.

    (%i) **tend:find_root(y(t),t,300,350);**   (%o) **312.19**

So the drop hits at $t = 312.2$ s. This is a *much* longer time than we found with linear resistance. This already indicates that the quadratic resistance has a greater effect on the drop's motion than did linear resistance, which suggests that quadratic resistance is the better model to use in this case. Now we can plot the velocity of the raindrop during its fall, using the expression for $v(t)$ from above. However, rather than plotting the velocity until the drop hits the ground we will plot the velocity of the drop during the first 10 s of its fall.

    (%i) **wxdraw2d(xlabel="t(s)",ylabel="v(m/s)",**
            explicit(v(t),t,0,10))$

Figure 2.10 shows that the raindrop speeds up as it begins to fall, but after only 2 s its speed levels out around 6 m/s. This suggests that the drop has rapidly reached its terminal speed and that the drop will fall with a constant speed from that point onward. This is consistent with what we saw earlier in the graph of y versus t. Now let's calculate the speed at impact.

    (%i) **v(tend);**   (%o) **−6.4156**

The drop has a speed of just under 6.42 m/s when it hits the ground. Is that speed the drop's terminal speed? To find the terminal speed we set the magnitude of the air

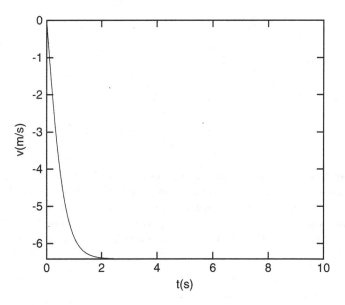

Fig. 2.10 Velocity of the falling raindrop (with quadratic resistance) as a function of time

resistance force equal to the drop's weight: $mg = cv^2$, which gives $v_t = \sqrt{mg/c}$. We evaluate this expression and compare it to the velocity on impact.

(%i) sqrt(m*g/c);   (%o) 6.4156

We see that the raindrop was indeed moving at its terminal speed of 6.42 m/s when it hit the ground. Note that the terminal speed for our water drop with quadratic resistance is much smaller than the terminal speed with linear resistance (128.6 m/s). Again, this suggests that the effects of quadratic resistance are much greater in this case than the effects of linear resistance. Using a quadratic model for air resistance gives more accurate results for the fall of our raindrop.

2.3.5 Projectile Motion with Quadratic Resistance

Now we examine projectile motion subject to quadratic resistance. This case is much more difficult, because the equations of motion are not *separable*, meaning they do not separate into equations that involve only x quantities and other equations that involve only y quantities. The reason for this is that the force of the air resistance is given by:

$$\vec{F}_{ar} = -cv^2\hat{v} = -cv\vec{v} = -c\sqrt{v_x^2 + v_y^2}\,\vec{v}. \qquad (2.22)$$

The x-component of this force involves both the x and y components of the object's velocity, etc. This is what prevents us from separating the equations.

The equations of motion are:

$$\dot{x} = v_x,$$

$$\dot{y} = v_y,$$

$$\dot{v}_x = -\frac{cv_x}{m}\sqrt{v_x^2 + v_y^2},$$

$$\dot{v}_y = -g - \frac{cv_y}{m}\sqrt{v_x^2 + v_y^2}. \tag{2.23}$$

We cannot derive an analytical solution for this set of non-separable equations. We can, however, examine the system's behavior by using numerical methods. The rk command uses a fourth order Runge–Kutta method for solving the system of ODEs. For more information about ODE algorithms and the Runge–Kutta method see Sects. 5.4 and A.3. To use rk we must first write the system of ODEs in the form

$$\frac{dy_1}{dx} = f(x, y_1, y_2, \ldots),$$

$$\frac{dy_2}{dx} = g(x, y_1, y_2, \ldots), \tag{2.24}$$

and so on. Fortunately, the system of ODEs in Eq. 2.23 is already in this form.

The basic rk command has four arguments, the first three of which are: a list of expressions for calculating the derivatives, a list of the functions for which we want to solve, and a list of initial values for each function. The final argument is a list that includes the integration variable, the initial and final values of the integration variable, and the step size. The rk command uses a fixed step size,[2] and it is important to keep this step size small so as to prevent significant error in the solution. The best way to ensure that you are using a sufficiently small step size is to run the calculation with smaller and smaller step sizes until you don't see any change in the solution. The output from rk is a list in which each element contains a value for the independent variable as well as the corresponding values for the functions for which we solved.

In the code below we apply rk to our example of a water drop fired from a water gun, but this time with quadratic resistance. Recall that the launch angle is 45° and the initial speed is 15 m/s. The code below generates a numerical solution to the ODEs in Eq. 2.23 using the appropriate parameters and initial conditions. We will try a step size of 0.1 s and find the solution from $t = 0$ to 3 s.

```
(%i) v0:15$ deg:%pi/180.0$ theta:45.0*deg$
     D:2e-3$ m:4.2e-6$ gamma:0.25$ c:gamma*D^2$ g:9.8$
     vx0:float(v0*cos(theta))$
     vy0:float(v0*sin(theta))$
```

---

[2]Some of the more sophisticated numerical ODE tools in *Maxima*, such as rkf45, use a variable step size. See Sect. A.3.

```
data:rk([vx,vy,-c*sqrt(vx^2+vy^2)*vx/m,-g-c*
   sqrt(vx^2+vy^2)*vy/m], [x,y,vx,vy],
   [0,0,vx0,vy0],[t,0,3,0.1] )$
```

The list of output from `rk` has been stored in an object named `data`. This object is, in fact, a list that consists of embedded lists. To get an idea of what the output looks like, look at one of the elements (the fifth) of this list.

(%i) **data[5];**   (%o)   [0.4, **2.693**, **2.0795**, **4.6921**, 1.8126]

As noted, this element is itself a list. The first element of this list is the value of the independent variable, $t = 0.4$ s. Recall that our step size is 0.1 s, so it makes sense that the fifth time in our list would be 0.4 s (the first time is 0, the second is 0.1 s, and so on). The next four values in this list are the values for x, y, v_x, and v_y, respectively, at $t = 0.4$ s. The `data` list contains one such list for each discrete time, starting at $t = 0$ and moving in steps of 0.1 s to $t = 3$ s. Therefore, `data` should have a total of 31 elements, as we can verify using *Maxima*'s `length` command.

(%i) **length(data);**   (%o)   **31**

We can now manipulate the `data` list to display our solution in various ways. For example, we can construct a list of ordered pairs (t, y) in order to make a plot of y versus t. To do this we can use `makelist`. The arguments of `makelist` are: an expression to calculate an element of the list, an index variable, the starting value of the index, and the ending value of the index.

We can generate our list of ordered pairs of t and y with `makelist` by selecting appropriate elements of `data` and organizing them into ordered pairs, with the index running from 1 to the number of elements in `data`. Then we can plot this list of points. Figure 2.11 shows the resulting plot.

```
(%i) yvt:makelist([data[i][1],data[i][3]],
        i,1,length(data))$ wxdraw2d(xlabel="t(s)",
        ylabel="y(m)", xaxis=true, point_size=0,
        points_joined=true,points(yvt))$
```

The plot indicates that the water drop hits the ground somewhere between $t = 1$ and 1.5 s. We would like to use `find_root` to find the precise time of impact, but to use `find_root` we must have a function, not a list of data points. We can obtain a function for this purpose by generating an *interpolating function* for our data points. One command for generating an interpolating function is `cspline`, which takes as its argument a list of ordered pairs and outputs an interpolating function. To use `cspline` we must first load the `interpol` package.

The code below shows how to load this package, use `cspline` to generate the interpolating function for our y versus t data, and then load the result into a new function $ys(t)$. Note: the output of `cspline` is in terms of "characteristic functions" which are not easy to interpret, so we do not display the output here. It is easier to load the result into a named function for later use. See Sect. A.4 for more information about `cspline` and other interpolation commands.

(%i)   **load(interpol)$**   **cspline(yvt)$ ys(x):="%$**

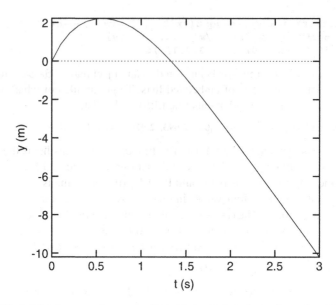

Fig. 2.11 The height of a water drop projectile (with quadratic resistance) as a function of time

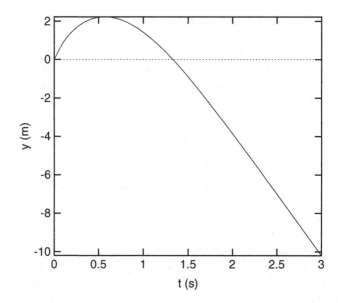

Fig. 2.12 Interpolating function for the height of a water drop projectile as a function of time

Now that we have our interpolating function, we can plot the function to compare it with our previous y versus t plot. Figure 2.12 shows the plot.

```
(%i) wxdraw2d(xlabel="t (s)",ylabel="y (m)",
     xaxis=true,explicit(ys(t),t,0,3) )$
```

The plot looks identical to our previous one, as it should. Now we can use our interpolating function to find the time of impact.

(%i) **tend:find_root(ys(t)=0,t,1,2);**   (%o)   **1.3291**

Our numerical solution indicates that the water drop hits the ground at $t = 1.3291$ s. It is instructive to redo the above analysis, but this time with a step size of 0.01 s. Try it! The new time of impact is $t = 1.3294$ s. Comparing these two values for the impact time shows us that our results for a step size of 0.1 s were very accurate, although the accuracy improves when the step size is reduced. Further reduction in step size will generate even more accurate results, but if we are satisfied with knowing the time of impact to the nearest millisecond then a step size of 0.1 s is sufficient.

To plot the trajectory, we need an interpolating function for $x(t)$ as well as $y(t)$. We proceed as before. We generate the list of x versus t points from the rk-generated list, data. Then we apply cspline to create the associated interpolating function and load the result into a new function $xs(t)$. Then we can create a parametric plot of y versus x, for both quadratic resistance and no air resistance. The resulting plot is shown in Fig. 2.13.

```
(%i) xvt:makelist([data[i][1],data[i][2]],i,1,
     length(data))$ cspline(xvt)$ xs(x):="%$
  wxdraw2d(xlabel="x (m)", ylabel="y (m)",
     user_preamble="set key center",key="Quadratic
     Resistance", parametric(xs(t),ys(t),t,0,tend),
     line_width=2, key="No Resistance",
     parametric(v0*cos(theta)*t,v0*sin(theta)*
     t-0.5*g*t^2,t,0,2*v0*sin(theta)/g))$
```

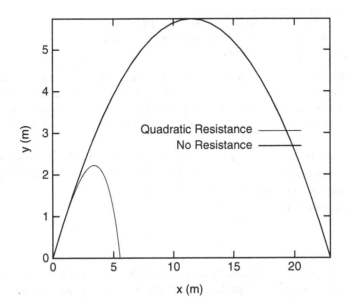

Fig. 2.13 Trajectory of the water drop with and without resistance

Quadratic resistance has a much greater effect on the drop than linear resistance did. While linear resistance reduced the range of the drop by a few meters, here we see that the range is reduced from about 23 m with no resistance to just over 5 m with quadratic resistance. We can determine the projectile's range with quadratic air resistance by evaluating the interpolating function $xs(t)$ at the landing time.

```
(%i)   xs(tend);          (%o)   5.5455
```

The water drop travels only 5.5 m before landing.

2.4   Charged Particles in an Electromagnetic Field

Another scenario we can explore is that of a charged particle moving through a static electromagnetic field. First we consider the case of the static electric field, with no magnetic field present. In this case we are free to choose our z-axis to align with the direction of the electric field, so $\vec{E} = E\hat{z}$. Since the force on a charged particle in an electric field is given by $\vec{F}_E = q\vec{E}$, where q is the particle's electric charge, the particle will accelerate along the z-axis. Newton's Second Law gives us

$$\ddot{z} = \frac{qE}{m}, \tag{2.25}$$

where m is the mass of the particle.

This differential equation is easy to solve by hand, but we can let *Maxima* do it for us.

```
(%i) atvalue(z(t),t=0,z0)$
     atvalue('diff(z(t),t),t=0,vz0)$
     soln:desolve(diff(z(t),t,2)=(q/m)*E,z(t));
(%o)  z(t) = q t² E / 2m + z0 + t vz0
```

The solution for $z(t)$ is quadratic in t with the linear term proportional to the initial z-component of velocity and the quadratic term proportional to the particle's charge and the strength of the electric field. The x- and y-components of the net force are both zero, so the charged particle will move with constant velocity along these axes.

Things get more interesting if we add in a magnetic field. This time we will assume that the magnetic field points in the z-direction (so $\vec{B} = B\hat{z}$ and the electric field can point in any direction (so $\vec{E} = E_x\hat{x} + E_y\hat{y} + E_z\hat{z}$). The magnetic force on the charged particle is

$$\vec{F}_B = q\vec{v} \times \vec{B}, \tag{2.26}$$

where $\vec{v}$ is the particle's velocity. We can use *Maxima* to evaluate the cross product. First we must load the vect package, and then use ~ to indicate a cross product.

```
(%i) load("vect")$              cross:[vx,vy,vz] ~ [0,0,B];
(%o)   −[0, 0, B]~[vx, vy, vz]
```

Note that *Maxima* does not fully evaluate the cross product. To get *Maxima* to evaluate the cross product we must use `express`.

```
(%i)  express(cross);      (%o)   [vy B, −vx B, 0]
```

Including both electric and magnetic forces in Newton's Second Law we find

$$\ddot{x} = (q/m)(B\dot{y} + E_x),$$
$$\ddot{y} = (q/m)(-B\dot{x} + E_y),$$
$$\ddot{z} = (q/m) + E_z. \tag{2.27}$$

We can solve this system of ODEs using `desolve`, after specifying the initial position (at the origin) and initial velocity components.

```
(%i) eq1:diff(x(t),t,2)=(q/m)*(B*diff(y(t),t)+Ex)$
     eq2:diff(y(t),t,2)=(q/m)*(-B*diff(x(t),t)+Ey)$
     eq3:diff(z(t),t,2)=(q/m)*Ez$
     atvalue(x(t),t=0,0)$
     atvalue(y(t),t=0,0)$
     atvalue(z(t),t=0,0)$
     atvalue('diff(x(t),t),t=0,vx0)$
     atvalue('diff(y(t),t),t=0,vy0)$
     atvalue('diff(z(t),t),t=0,vz0)$
soln:desolve([eq1,eq2,eq3],[x(t),y(t),z(t)]);
(%o)
```

$$x(t) = \frac{(m\,vx0\,B - Ey\,m)\sin\left(\frac{q\,t\,B}{m}\right) + (-m\,vy0\,B - Ex\,m)\cos\left(\frac{q\,t\,B}{m}\right) + (m\,vy0 + Ey\,q\,t)\,B + Ex\,m}{q\,B^2}$$

$$y(t) = \frac{(m\,vy0\,B + Ex\,m)\sin\left(\frac{q\,t\,B}{m}\right) + (m\,vx0\,B - Ey\,m)\cos\left(\frac{q\,t\,B}{m}\right) + (-m\,vx0 - Ex\,q\,t)\,B + Ey\,m}{q\,B^2}$$

$$z(t) = t\,vz0 + \frac{Ez\,q\,t^2}{2\,m}$$

Maxima asks us whether the quantity *Bmq* is zero or nonzero. The result shown above is only valid for nonzero magnetic field. We can rewrite this result in much simpler form by defining a few constants.

$$\omega = \frac{Bq}{m},$$
$$\alpha = \frac{m(Bv_{x0} - E_y)}{B^2 q}, \tag{2.28}$$
$$\beta = \frac{m(Bv_{y0} + E_x)}{B^2 q}.$$

Then

$$x(t) = \alpha \sin(\omega t) + \beta(1 - \cos(\omega t)) + (E_y/B)t,$$
$$y(t) = \beta \sin(\omega t) - \alpha(1 - \cos(\omega t)) - (E_x/B)t, \tag{2.29}$$
$$z(t) = v_{z0}t + E_z q t^2/(2m).$$

Fig. 2.14 Trajectory of a particle moving in a static magnetic field. Units are Bohr radii, a_0

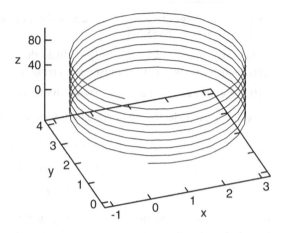

We examine this solution for the case where there is no electric field. Suppose that an electron moves through a magnetic field. In atomic units the mass the electron is 1, its charge is -1 and distances are measured in units of the Bohr radius $a_0 \approx 5.3 \times 10^{-11}$ m. Velocity is measured in units of the speed of light times the fine structure constant, or about 2.2×10^6 m/s. The atomic unit for magnetic field is approximately 2.35×10^9 G. The code below defines the solution for this case (with specific values for the magnetic field and initial velocity components) and generates the 3D parametric plot of the motion shown in Fig. 2.14.

```
(%i) q:-1$ m:1$ B:1$ Ex:0$ Ey:0$ Ez:0$
     vx0:2$ vy0:-1$ vz0:2$ omega:B*q/m$
     alpha:m*(B*vx0-Ey)/(B^2*q)$
     beta:m*(B*vy0+Ex)/(B^2*q)$
     x(t):=alpha*sin(omega*t)+beta*(1-cos(omega*t))+
       (Ey/B)*t$
     y(t):=beta*sin(omega*t)-alpha*(1-cos(omega*t))-
       (Ex/B)*t$
     z(t):=vz0*t+Ez*q*t^2/(2*m)$
     wxdraw3d(nticks=200,parametric(x(t),y(t),
       z(t),t,0,50), xlabel="x",ylabel="y",zlabel="z",
       view=[45,340], xtics=1,ytics=1,ztics=40)$
```

Note the use of the `nticks` option within wxdraw3d. This option sets the number of sampling points used to generate the plot. Higher values lead to a smoother plot. The motion of the charged particle is along a helical path, with the axis of the helix aligned with the direction of the magnetic field (the z-direction, in this case). But what, we might wonder, determines the radius of the helix?

To answer that question we can consider the case of a particle moving in a static magnetic field, with no electric field. If the particle's initial velocity is in the x–y plane, then the magnetic force on the particle will also be in the x–y plane and thus there will be no motion in the z-direction. We can show that the motion of the particle will be in a circle within the x–y plane. We can simplify Eq. 2.29 for this case by setting the electric field components to zero. We can then use *Maxima* to show that $x(t)$ and $y(t)$ satisfy the equation for a circle of radius R centered at $(\beta, -\alpha)$:

$$(x(t) - \beta)^2 + (y(t) - \alpha)^2 = R^2. \tag{2.30}$$

The code below defines $x(t)$ and $y(t)$ according to Eq. 2.29 (with no electric field), evaluates the left side of Eq. 2.30, and then simplifies the result using `factor` and `trigreduce`.

```
(%i)  kill(functions, values)$
      omega:B*q/m$ alpha:m*vx0/(B*q)$
      beta:m*vy0/(B*q)$
      x(t):=alpha*sin(omega*t)+beta*(1-cos(omega*t))$
      y(t):=beta*sin(omega*t)-alpha*(1-cos(omega*t))$
      lhs:(x(t)-beta)^2+(y(t)+alpha)^2;
```

$$(\%o) \quad \left(\frac{m \, vy0 \sin\left(\frac{q t B}{m}\right)}{q B} - \frac{m \, vx0 \left(1 - \cos\left(\frac{q t B}{m}\right)\right)}{q B} + \frac{m \, vx0}{q B} \right)^2 +$$

$$\left(\frac{m \, vx0 \sin\left(\frac{q t B}{m}\right)}{q B} + \frac{m \, vy0 \left(1 - \cos\left(\frac{q t B}{m}\right)\right)}{q B} - \frac{m \, vy0}{q B} \right)^2$$

```
(%i)  trigreduce(factor(lhs));
```
$$(\%o) \quad \frac{m^2 \, vy0^2}{q^2 B^2} + \frac{m^2 \, vx0^2}{q^2 B^2}$$

The result is a constant, showing that Eq. 2.30 is satisfied if

$$R = mv_0/(Bq), \tag{2.31}$$

where $v_0 = \sqrt{v_{x0}^2 + v_{y0}^2}$ is the initial speed of the particle. So the radius is determined by the speed of the particle, the strength of the magnetic field, and the charge-to-mass ratio of the particle. The code below generates a plot of the path of a particle with $q = 2$, $m = 5$, $B = 1$, $v_0 = 5$ in atomic units. The plot is shown in Fig. 2.15. (Note the use of `proportional_axes=xy` to force the plot to have a square aspect ratio.) It is easy to see that the particle follows a circular path with a radius of $R = 12.5$, centered on $(-7.5, -10)$ in Bohr radii.

```
(%i)  q:2$ m:5$ B:1$ vx0:4$ vy0:-3$
      wxdraw2d(nticks=50,proportional_axes=xy,
        parametric(x(t),y(t),t,0,20),
        xlabel="x",ylabel="y")$
```

If both magnetic and electric fields are present, we can arrange a situation where the magnetic and electric forces exactly cancel each other. For this situation to occur we must have

$$\vec{F}_B = -\vec{F}_E \rightarrow q\vec{v} \times \vec{B} = -q\vec{E} \rightarrow \vec{v} \times \vec{B} = -\vec{E}. \tag{2.32}$$

The electric field must be perpendicular to both the magnetic field and the particle's velocity. For example, suppose the magnetic field is in the $\hat{z}$ direction, while the electric field is in the $\hat{x}$ direction. The magnetic and electric forces will

Fig. 2.15 Circular trajectory
of a charged particle in a
static magnetic field. Units
are Bohr radii, a_0

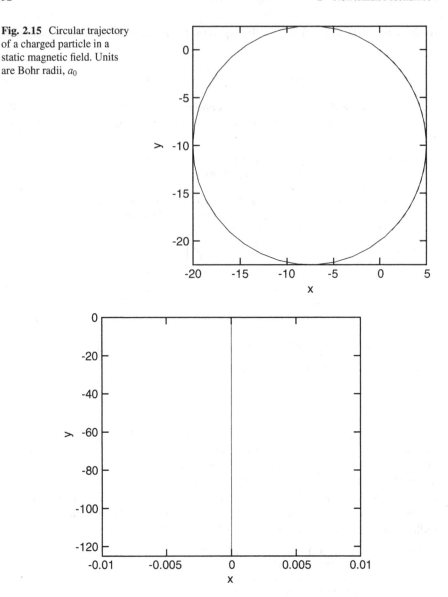

Fig. 2.16 Trajectory illustrating the cancelation of electric and magnetic forces. Units are Bohr
radii, a_0

cancel if the particle's initial velocity is in the y-z plane with $v_{y0} = -E/B$ (the value
of v_{z0} does not affect the cancellation of forces). This behavior is illustrated below
for the case $q = 2$, $m = 4$, $B = 2$, $E = 5$, $v_{y0} = -2.5$ and $v_{z0} = 0$ in atomic units.
The plot in Fig. 2.16 shows the trajectory of this particle, which is simply a straight
line.

```
(%i)  q:2$ m:4$ B:2$ Ex:5$ Ey:0$ Ez:0$ vx0:0$
      vy0:-2.5$ vz0:0$    omega:B*q/m$
      alpha:m*(B*vx0-Ey)/(B^2*q)$
      beta:m*(B*vy0+Ex)/(B^2*q)$
      x(t):=alpha*sin(omega*t)+beta*(1-cos(omega*t))+
        (Ey/B)*t$
      y(t):=beta*sin(omega* t)-alpha*(1-cos(omega*t))-
        (Ex/B)*t$
      wxdraw2d(parametric(x(t),y(t),t,0,50),
        xlabel="x",ylabel="y")$
```

Finally, we consider a more general case of motion in combined electric and magnetic fields. The path shown in Fig. 2.17, which is generated by the code below, illustrates a case where the electric and magnetic forces do not cancel. The path is helical, but with the helix curving downward (in the $-\hat{z}$ direction) and spreading out.

```
(%i)  q:2$ m:5$ B:4$ Ex:2$ Ey:-1$ Ez:-1$
      vx0:1$ vy0:-1$ vz0:2$
      omega:B*q/m$ alpha:m*(B*vx0-Ey)/(B^2*q)$
      beta:m*(B*vy0+Ex)/(B^2*q)$
      x(t):=alpha*sin(omega*t)+
        beta*(1-cos(omega*t))+(Ey/B)*t$
      y(t):=beta*sin(omega*t)-alpha*(1-cos(omega*t))-
        (Ex/B)*t$
      z(t):=vz0*t+Ez*q*t^2/(2*m)$
      wxdraw3d(view=[20,160],nticks=100,
        parametric(x(t),y(t),z(t),t,0,20),xlabel="x",
        ylabel="y",zlabel="z",xtics=1,ytics=2,ztics=20);
```

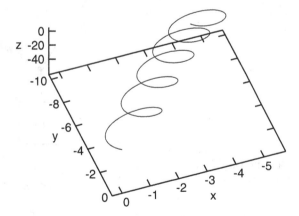

Fig. 2.17 Trajectory of a charged particle in static electric and magnetic fields. Units are Bohr radii, a_0

2.5   Exercises

1. Consider the arrangement shown in Fig. 2.18, which consists of a mass m hanging from a (massless) wire. The wire is attached to the end of a uniform, rigid rod of mass M and length L. That end of the rod is also attached to another wire with tension T that attaches at an angle ϕ to a horizontal surface. The opposite end of the rod attaches at an angle θ to a hinge on the same horizontal surface. Determine the tension T in the angled wire, as well as the x- and y-components of the force that the hinge exerts on the rod, as functions of θ and ϕ. Evaluate the solution in the limit $\theta \to \pi/2$. Also evaluate the solution in the limit $\phi \to \pi/2$. Do your answers for the two limiting cases make sense?

2. Consider two blocks stacked on top of each other and sitting on an inclined plane, as shown in Fig. 2.19. A force of magnitude F is applied to the top block in a direction perpendicular to the surface of the plane and up the slope. There is friction between the plane and block 1, resulting in a force of magnitude $f_k = \mu_k N_1$, where N_1 is the normal force exerted on block 1 by the plane and μ_k is the coefficient of kinetic friction. There is also a force of static friction, with magnitude f_s, between the two blocks. If the blocks are pulled up the slope at constant speed, determine the applied force F and the force of static friction f_s, as well as the normal forces on the two blocks. Recall that $f_s/N_2 \le \mu_s$, where N_2 is the normal force exerted on the top block by the bottom block and μ_s is the coefficient of static friction between the blocks. What is the minimum value of μ_s such that the blocks can be pulled up the slope without the top block slipping off?

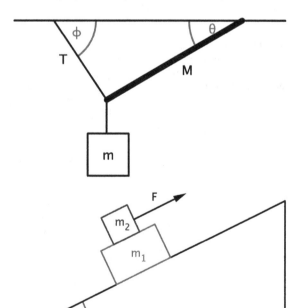

Fig. 2.18 Diagram for Exercise 1

Fig. 2.19 Diagram for Exercise 2

3. According to a later account by his student Viviani, Galileo performed a demonstration in which he dropped two metal balls from the top of the Leaning Tower of Pisa. This demonstration, performed sometime around 1590 according to Viviani, was intended to show that bodies composed of the same material will fall at the same rate regardless of their mass. This idea of Galileo's was in direct contradiction to Aristotle's physics, which stated that falling bodies fall at a rate that is proportional to their mass.

   Suppose Galileo dropped two balls of solid lead, one with a diameter of 20 cm and another with a diameter of 10 cm, from the top of the 56 m high tower. The density of lead is 11,342 kg/m^3. Determine the mass of each ball. Should you assume that air resistance is linear in this case? Quadratic? Explain the reasons for your choice. Use the solution for falling motion, with the appropriate model of air resistance, to determine the time at which each of the two balls hits the ground. Is there a difference? Which ball hits first? When the first ball hits, how far is the other ball above the ground?

4. Later investigators carried out a new version of the falling balls experiment (see previous question) using spheres of the same size, but composed of different materials. Repeat parts the previous question for the case where both spheres have a diameter of 10 cm, but one is made of lead and the other of oak wood (density: 740 kg/m^3). In which case (two lead spheres, or one lead and one wood sphere) is the difference in time of fall, which is due solely to air resistance, more noticeable?

5. Consider the case of an object falling from the height of the International Space Station (see Exercises 1. and 2. from Chap. 1). Assume the object is a bowling ball with a mass of 6.8 kg and a diameter of 21.6 cm. Determine how much time passes between the moment the ball is dropped (from rest) and the moment it hits the Earth, as well as the speed with which the ball hits the Earth. This time use a model in which there is no air resistance at all above the troposphere (which extends from Earth's surface up to a height of 17 km). Within the troposphere the ball will be subject to air resistance according to the models of air resistance we have been considering. You may wish to break the problem into two parts (one for the ball falling to the top of the troposphere, and then another for the ball falling from the top of the troposphere to the surface of Earth). Be careful with initial conditions for the second part. Think carefully about which model of air resistance you want to use (linear, quadratic, or both together). Make sure you justify your choice.

6. Recent research from the National Baseball Hall of Fame suggests that the longest home run ever hit in a Major League Baseball game was belted by none other than Babe Ruth, on 18 July 1921 at Briggs Stadium in Detroit, MI. The ball apparently traveled 575 ft (175.3 m) in the air before landing. We will assume that Ruth hit the ball at a height of 1 m above ground. The mass of a baseball is 0.145 kg and the diameter of the baseball is 7.4 cm. Can we determine how fast the ball was going when it came off of Ruth's bat?

   The following questions will guide you through the process of determining how hard Ruth hit that historic home run.

(a) To begin with, assume that Ruth hit the ball at the perfect angle. This would be the angle that maximizes the distance the ball travels before landing. If there were no air resistance, then this would be 45° above horizontal. But with air resistance this result no longer holds, so you must first determine this optimal angle. Use rk to solve the equations of motion for the baseball, assuming it was launched with a speed of 100 mph (44.7 m/s). Find the distance (d) the ball travels using several different launch angles (θ). You may want to focus on angles between 30° and 50°. Compile your results into a list of ordered pairs (θ, d). Use an interpolating function to find the launch angle that maximizes the distance (see Sect. A.4).

(b) Once you have determined the optimal launch angle, use rk to solve the equations of motion for various launch speeds. Experiment with the value for the initial speed until you find a speed that makes the ball travel 175.3 m before it lands. Report your result in both m/s and mph.

7. In Sect. 2.4 we found that a charged particle moving through perpendicular electric and magnetic fields ($\vec{B} = B\hat{z}$ and $\vec{E} = E\hat{x}$) will experience no net electromagnetic force if that particle's velocity is $\vec{v} = -(E/B)\hat{y}$. Create a plot illustrating the paths of three particles moving through this electromagnetic field. All three particles start at the origin and have initial velocities in the $-\hat{y}$ direction. One has initial speed E/B, another has a somewhat greater initial speed, and the third has a somewhat smaller initial speed. Explain how this setup could be used as a "velocity selector" that separates out particles of a specific velocity, from a beam of particles traveling at a variety of different velocities.

8. In Sect. 2.4 we found that a charged particle moving perpendicular to a static magnetic field ($\vec{B} = B\hat{z}$) will move in a circle of radius $R = mv_0/(Bq)$, where v_0 is the particle's speed and m/q is the particle's mass-to-charge ratio. Create a plot illustrating the paths of three particles moving through this magnetic field. All three particles start at the origin and have the same initial velocity in the $\hat{x}$ direction (perhaps as a result of passing through the velocity selector discussed in the previous problem), but each particle has a different mass-to-charge ratio. Explain how this setup could be used with particle detectors to analyze the chemical composition of a beam of singly ionized atoms.

Chapter 3
Momentum and Energy

Although Newton's Laws of Motion form the fundamental basis of Newtonian Mechanics, there are several other important concepts that are an integral part of that subject. In this chapter we use *Maxima* to examine some of these concepts, including linear and angular momentum, center of mass, work, and energy.

3.1 Collisions: Conservation of Momentum

The momentum $\vec{p}$ of an object is defined as the product of the object's mass and its velocity:

$$\vec{p} = m\vec{v}. \tag{3.1}$$

We can write Newton's Second Law as

$$\vec{F}_{\text{net}} = \frac{d\vec{p}}{dt}. \tag{3.2}$$

If the mass of the object is constant then Eq. 3.2 reduces to the more familiar $\vec{F}_{\text{net}} = m\vec{a}$. Equation 3.2 shows us that what forces do, ultimately, is change the momentum of objects.

Newton's Third Law tells us that forces are interactions between two objects. Forces involve an exchange of momentum between the two bodies. Any momentum gained by one body must be lost by the other.[1] Therefore, the total momentum of any isolated system of bodies (bodies that interact with each other, but not with anything else) must remain constant.

---

[1]Here we are ignoring electric and magnetic fields, which can also exchange momentum with bodies and other fields.

© Todd Keene Timberlake & J. Wilson Mixon, Jr. 2016
T.K. Timberlake, J.W. Mixon, *Classical Mechanics with Maxima*, Undergraduate
Lecture Notes in Physics, DOI 10.1007/978-1-4939-3207-8_3

As a specific example, consider a head-on collision between two particles with masses m_1 and m_2. The motion before and after the collision must take place along the line of initial approach. We need not worry about the vector nature of momentum and velocity except to pay attention to the signs of these quantities. The total momentum of the system before the collision is $m_1 v_{10} + m_2 v_{20}$, where v_{10} is the initial velocity of the particle with mass m_1 and v_{20} is the initial velocity of the particle with mass m_2. The total momentum after the collision is $m_1 v_1 + m_2 v_2$ where v_1 and v_2 are the final velocities of the two particles. Conservation of momentum tells us that

$$m_1 v_{10} + m_2 v_{20} = m_1 v_1 + m_2 v_2. \tag{3.3}$$

If we know the velocities before the collision, then Eq. 3.3 can help us to determine the velocities after the collision. But Eq. 3.3 is not enough, because it is only a single equation with two unknowns. We need more information. Conservation of momentum alone is not sufficient to determine the outcome of this collision. Further information comes from the nature of the collision that is being examined.

Let us focus on some special cases of this head-on collision. First, we consider a *totally inelastic collision* in which the two particles stick together after the collision. In that case, $v_1 = v_2 = v$, so Eq. 3.3 has only one unknown. We can solve the equation using *Maxima*.

```
(%i)  eq1:m1*v10+m2*v20 = m1*v + m2*v$
      solnI : solve(eq1, v );
(%o)  [v = m2 v20+m1 v10 / m2+m1]
```

This result for v tells us the velocity of both particles after the collision. This collision is inelastic, which means that some of the kinetic energy in the system is lost in the collision.[2] The kinetic energy of a particle is given by $K = (1/2)mv^2$ where m is the particle's mass and v is its speed. We can calculate how much kinetic energy is lost in this collision by subtracting the final kinetic energy from the initial kinetic energy. The code below shows how to calculate the kinetic energy lost and then use `factor` to simplify the result.

```
(%i)  KElost:(1/2)*(m1*v10^2+m2*v20^2-(m1+m2)*
      ((m2*v20+m1*v10)/(m2+m1))^2);
(%o)  (-m2-m1)(m2 v20+m1 v10)^2/(m2+m1)^2 +m2 v20^2+m1 v10^2 / 2

(%i)  factor(KElost);        (%o)  m1 m2 (v20-v10)^2 / 2(m2+m1)
```

We can rewrite this result as

$$K_{\text{lost}} = \frac{1}{2}\mu v_{r0}^2, \tag{3.4}$$

---

[2]In fact, this kind of collision loses the maximum amount of kinetic energy consistent with conservation of momentum.

where μ is the reduced mass of the system,

$$\mu = \frac{m_1 m_2}{m_1 + m_2}, \tag{3.5}$$

and $v_{r0} = v_{10} - v_{20}$ is the relative velocity of the two particles before the collision. The relative velocity of the two particles is zero after the collision, since they are stuck together. The greater the initial relative speed, and thus the greater the change in the relative speed during the collision, the more kinetic energy will be lost.

What if no kinetic energy is lost in the collision? This type of collision is called *elastic*, and in this case the equality of the initial and final kinetic energies gives us a second equation to go along with Eq. 3.3:

$$\frac{1}{2} m_1 v_{10}^2 + \frac{1}{2} m_2 v_{20}^2 = \frac{1}{2} m_1 v_1^2 + \frac{1}{2} m_2 v_2^2. \tag{3.6}$$

We now have a system of two equations with two unknowns, which we can solve.

```
(%i) eq1:m1*v10 + m2*v20 = m1*v1 + m2*v2$
     eq2: (1/2)*m1*v10^2+(1/2)*m2*v20^2=(1/2)*m1*v1^2+
     (1/2)*m2*v2^2$
     soln : solve([eq1, eq2], [v1,v2] );
```
$$(\%o) \quad [[v1 = v10, \quad v2 = v20],$$
$$[v1 = \frac{2\,m2\,v20 + (m1 - m2)\,v10}{m2 + m1}, \quad v2 = \frac{(m2 - m1)\,v20 + 2\,m1\,v10}{m2 + m1}]]$$

There are two different solutions. The first solution is the trivial case in which the particles do not interact and thus their velocities do not change. We are interested in the second solution, in which the particles actually collide and change their velocities. Let's examine the relative velocity of the particles after the collision. We can load the right-hand sides of the nontrivial solutions into new variables, calculate the difference between those two variables, and then simplify the result using `factor`.

```
(%i) [v1,v2]:[rhs(soln[2][1]), rhs(soln[2][2])];
```
$$(\%o) \quad [\frac{2\,m2\,v20 + (m1 - m2)\,v10}{m2 + m1}, \quad \frac{(m2 - m1)\,v20 + 2\,m1\,v10}{m2 + m1}]$$

```
(%i)  factor(v1-v2);        (%o)  v20 - v10
```

The relative velocity before the collision is $v_{10} - v_{20}$, and the relative velocity after the collision is $v_{20} - v_{10}$. The magnitude of the relative velocity is unchanged by this collision, only the sign changes. This is consistent with the result for the totally inelastic collision, in which the amount of kinetic energy lost depends on how much the relative speed changes. In an elastic collision the relative speed does not change at all and no kinetic energy is lost.

We now examine elastic collisions further by addressing some limiting cases. What happens if one of the masses is at rest before the collision? We can compute the limit of our solutions as $v_{20} \to 0$.

```
(%i) [v1s,v2s]:limit([v1,v2],v20,0);
```
$$(\%o) \quad [-\frac{(m2 - m1)\,v10}{m2 + m1}, \quad \frac{2\,m1\,v10}{m2 + m1}]$$

This solution implies that the sign of v_2 is always positive, so the particle that was stationary before the collision will end up moving in the same direction that the incident particle was moving before the collision. However, the sign v_1 depends on the relative masses of the two particles. If $m_1 > m_2$ then the incident particle will continue moving in the same direction after the collision. If $m_1 < m_2$ then the incident particle will change its direction of motion.

We can take a look at some further limiting cases. What if one particle is stationary but it has a very large mass? We can examine this case by finding the limit as the mass of the stationary particle becomes infinite, or by finding the limit as the mass of the initially moving particle goes to zero.

```
(%i)  [limit([v1s,v2s],m2,inf),limit([v1s,v2s],m1,0)];
(%o)   [[-v10, 0], [-v10, 0]]
```

We see that both limits produce the same result. In this case the incident particle bounces off of the massive stationary particle and ends up going back the way it came at the same speed. What if the stationary particle has a very small mass? We can examine the limit as $m_1 \to \infty$ or the limit as $m_2 \to 0$ to find out what happens in this case.

```
(%i)  [limit([v1s,v2s],m1,inf),limit([v1s,v2s],m2,0)];
(%o)   [[v10, 2 v10], [v10, 2 v10]]
```

Again, both limits give the same result. In this case the speed of the incident particle is unchanged, while the initially stationary particle moves off in the same direction at twice the speed of the incident particle. To examine the more general case of a collision where one mass is initially stationary we can plot the final velocities of the two particles (stated as fractions of the initial speed of the incident particle) as a function of the mass ratio m_1/m_2.

The code to produce such a plot is shown below. Figure 3.1 shows the resulting plot, which illustrates how the velocity of each particle after the collision (in units of the incident particle's initial speed) depends on the mass ratio for the two particles.

```
(%i)  x_max:50$
      [v1s(x):=(x-1)/(1+x), v2s(x):=2*x/(1+x)]$
      wxdraw2d( xaxis = true,xlabel="m1/m2",
        ylabel="velocity/v0", key="v1",
        explicit( v1s(x), x, 0, x_max ), line_width=3,
        key = "v2", explicit( v2s(x), x, 0, x_max),
        yrange=[-1,2.5],user_preamble="set key left")$
```

Figure 3.1 matches the results for $m_1/m_2 \to 0$ and $m_1/m_2 \to \infty$ that we found earlier. It also graphically illustrates the fact that the relative velocity does not change, because the curves remain separated by one unit (i.e., by v_0) for all mass ratios. We can zoom in on the region of small mass ratios by changing the value of xmax in the code above. Figure 3.2 shows the resulting plot for xmax equal to 5. A close inspection of this plot will show that when $m_1 = m_2$ the incident particle stops ($v_1 = 0$) while the initially stationary particle moves forward at a velocity equal to the incident particle's velocity before the collision ($v_2 = v_0$).

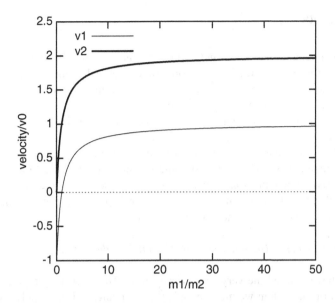

Fig. 3.1 Velocities of the particles after a head-on collision in which particle 2 was initially stationary. The two curves show the velocity for each particle (relative to the initial speed of the moving particle) as a function of the mass ration m_1/m_2

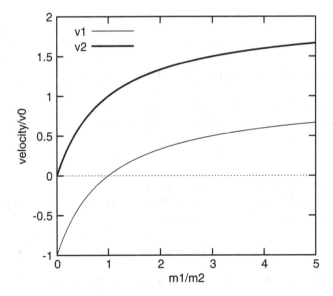

Fig. 3.2 Detail of Fig. 3.1 over a smaller range of mass ratios

3.2   Rockets

A rocket is a vehicle that burns fuel and expels the exhaust. The expelled exhaust gains momentum in one direction, so in order to conserve the total momentum of the rocket–exhaust system the rocket must gain momentum in the opposite direction. We can work out the details of this process to better understand how rockets are propelled. To simplify our analysis, we assume that the rocket is moving in one dimension only.

Suppose that at time t the rocket (and its fuel) has mass m and velocity v, and thus its momentum is $p(t) = mv$. At some (infinitesimally) later time $t + dt$ the rocket will have a new mass $m + dm$, where $dm < 0$ because some of the fuel that composed the rocket's initial mass has now been expelled as exhaust, and a new velocity $v + dv$. The rocket's new momentum is

$$p_r(t + dt) = (m + dm)(v + dv) \approx mv + v\,dm + m\,dv, \qquad (3.7)$$

where we have ignored the very small $dm\,dv$ term. However, to compute the total momentum of the system we must include the momentum of the expelled exhaust. The mass of this exhaust will be $-dm$ and the velocity will be $v - v_{ex}$, where v_{ex} is the velocity at which the exhaust is expelled relative to the rocket. So the momentum of the exhaust is

$$p_e(t + dt) = -v\,dm + v_{ex}\,dm. \qquad (3.8)$$

The total change in momentum of the system from time t to time $t + dt$ is thus

$$dp = p_r(t + dt) + p_e(t + dt) - p(t) = m\,dv + v_{ex}\,dm. \qquad (3.9)$$

If the net external force on the system is F_{ext} then Newton's Second Law tells us that the change in momentum must be $dp = F_{ext}dt$, and therefore

$$m\,dv + v_{ex}\,dm = F_{ext}\,dt. \qquad (3.10)$$

We will refer to Eq. 3.10 as the rocket equation.

We first consider a rocket moving in deep space, not subject to any external forces. In this case Eq. 3.10 simplifies to

$$m\,dv + v_{ex}\,dm = 0, \qquad (3.11)$$

and we can rearrange the equation in order to separate the variables:

$$dv = -\frac{v_{ex}\,dm}{m}. \qquad (3.12)$$

We can rewrite this equation as an integral equation,

$$v_f - v_0 = -\int_{m_0}^{m_f} \frac{v_{ex}}{m} dm. \tag{3.13}$$

Now we use *Maxima* to evaluate this integral, keeping in mind that $m_f - m_0 < 0$ since the rocket burns fuel and thus loses mass as it moves.

```
(%i)  assume(mf>0,m0>0,mf-m0>0)$
      integrate (-vex/m,m,m0,mf);
(%o)  − (log (mf) − log (m0)) vex
```

Properties of logarithms allow us to rewrite this result as

$$v_f - v_0 = v_{ex} \ln(m_0/m_f). \tag{3.14}$$

Suppose our rocket is a Saturn V that is floating at rest in deep space with an initial mass of 2.8×10^6 kg. The rocket's S-IC engine expels exhaust at a speed of 2500 m/s. If the engine fires for 2 min and 40 s, burning 2.2×10^6 kg of fuel, how fast will rocket be going at the end of the burn? We can compute the final mass, $m_f = 2.8 \times 10^6 - 2.2 \times 10^6 = 6 \times 10^5$ kg, and then substitute values into Eq. 3.14.

```
(%i)  vli:0$ vlex:2500$ m0:2.8e6$ mf:6e5$
      vlf:vli+vlex*log(m0/mf);
(%o)  3851.1
```

This rocket will be moving at a speed of over 3850 m/s by the end of the burn. What if the rocket was on Earth and was being launched upward against gravity (but with no air resistance)? In that case we have an external force, $F_{ext} = mg$ (assuming that the rocket stays close enough to Earth to treat the gravitational field as constant in magnitude). Inserting this external force into the rocket equation (Eq. 3.10), dividing by dt and rearranging the terms gives us

$$\frac{dv}{dt} = -g - \frac{v_{ex}}{m}\frac{dm}{dt}. \tag{3.15}$$

If the rocket engine burns fuel at a constant rate k, then $dm/dt = -k$ (which is negative because the rocket is *losing* mass) and $m = m_0 - kt$. Our equation then becomes

$$\frac{dv}{dt} = -g + \frac{v_{ex}k}{m_0 - kt}. \tag{3.16}$$

We can integrate both sides of this equation. The left side obviously gives $v(t) - v_0$. We can use *Maxima* to integrate the right side.

```
(%i)  kill(values)$ integrate(-g+vex*k/(m0-k*t),t);
(%o)  −log (m0 − k t) vex − g t
```

If we add the constant v_{ex} In m_0 to this result to ensure that $v(0) = v_0$ we find

$$v(t) = v_0 - v_{ex} \ln(\frac{m_0}{m_0 - kt}) - gt = \frac{dy}{dt}, \tag{3.17}$$

where y represents the height of the rocket above ground. Integrating once again yields $y(t)$, up to an additive constant.

```
(%i) v(t):=v0 + vex*log(m0/(m0 - k*t)) - g*t$
     integrate(v(t),t)$        ys(t):="%;
```
$$(\%o) \ ys(t) := \left(t\log\left(\frac{m0}{m0-kt}\right) - k\left(-\frac{m0\log(kt-m0)}{k^2} - \frac{t}{k}\right)\right) vex + t\,v0 - \frac{gt^2}{2}$$

The additive constant is determined by the fact that we want $y(0) = y_0$, so $y(t) = ys(t) - ys(0) + y_0$. We can construct the correct $y(t)$ function in *Maxima*.

```
(%i) y(t):=ys(t)-ys(0)+y0$        y(t);
```
$$(\%o) \ y0 + \left(t\log\left(\frac{m0}{m0-kt}\right) - k\left(-\frac{m0\log(kt-m0)}{k^2} - \frac{t}{k}\right)\right) vex -$$
$$\frac{m0\log(-m0)\,vex}{k} + t\,v0 - \frac{gt^2}{2}$$

This result can be simplified using *Maxima*'s expand command.

```
(%i)  expand(y(t));
```
$$(\%o) \ y0 + t\log\left(\frac{m0}{m0-kt}\right) vex + \frac{m0\log(kt-m0)\,vex}{k} + t\,vex - \frac{m0\log(-m0)\,vex}{k} +$$
$$t\,v0 - \frac{gt^2}{2}$$

Now we determine what would happen if our Saturn V rocket were launched from Earth's surface. First we need to determine the fuel burn rate, $k = (m_0 - m_f)/t$. Using the values given above we can calculate k.

```
(%i) m0:2.8e6$ mf:6e5$ tf:160$ k:(m0-mf)/tf;
(%o)    13750.
```

So the S-IC engine burns fuel at a rate of 13,750 kg/s. Now we can compute the velocity of the rocket at the end of the burn.

```
(%i) vex:2500$ g:9.8$ v0:0$ v(tf);        (%o) 2283.1
```

In this case the rocket is traveling at a speed of 2283 m/s at the end of the burn, much slower than when the rocket was in deep space. That makes sense because gravity is pushing down on the rocket and slows its ascent. We can use the solution for $y(t)$ to calculate the height of the rocket above Earth's surface at the end of the burn.

```
(%i) y(t):= vex*t-(vex/k)*(m0-k*t)*log(m0/(m0-k*t))+
     v0*t-g*t^2/2$ vex:2500$ g:9.8$ v0:0$ m0:2.8e6$
     mf:6e5$ tf:160$ k:(m0-mf)/tf$ yf:y(tf);
(%o)    106511.
```

In this case, the rocket will reach a height of more than 106 km before the fuel in the S-IC engine is spent. Although this height is still small compared to the radius of Earth (6371 km), it is greater than 1 % of that radius and we might worry that our assumption of constant gravity will lead to errors. More importantly, a real rocket must travel through Earth's atmosphere and will be subject to air resistance. We get a better idea about the real motion of a Saturn V by using an external force that incorporates universal gravitation and quadratic air resistance:

$$F_{\text{ext}} = -\frac{GMm}{(R+y)^2} - cv^2,$$ (3.18)

where R is the radius of Earth, $c = \gamma D^2$, D is the diameter of the rocket, and $\gamma \approx 0.25\,\text{N s}^2/\text{m}^4$.[3] Substituting this external force into our rocket equation (Eq. 3.10) and solving for dv/dt we find

$$\frac{dv}{dt} = -\frac{GM}{(R+y)^2} + \frac{kv_{\text{ex}} - cv^2}{m_0 - kt}.$$ (3.19)

Once we specify all of our parameters, we can solve Eq. 3.19 numerically using rk with a time step of 0.1 s.

```
(%i) D:10.1$ gamma:0.25$ c:gamma*D^2$ m0:2.8e6$
     GM:3.99e14$ R:6.371e6$ vex:2500$ tf:160$
     mf:6e5$ k:(m0-mf)/tf$
     data1: rk([v,-GM/(R+y)^2 + (k*vex-c*v^2)/
       (m0-k*t)],[y,v], [0,0],[t,0,tf,0.25])$
```

To find out the rocket's velocity and position at the end of the burn we only need to look at the last element in the data list produced by rk.

```
(%i) length(data1);
     data1[length(data1)];
(%o) 641    [[160.0,72144.,1003.6]
```

This shows us that the rocket reaches a height of over 72 km. But the Earth's troposphere only extends about 17 km above the surface. If we assume that the rocket experiences no resistance after it leaves the troposphere then we need to determine the rocket's velocity at the top of the troposphere and then continue our solution from that point without air resistance. To determine the time at which the rocket reaches the top of the troposphere we first fit the y versus t data from rk using cspline. We can then plot the resulting fit function to find the approximate time at which $y = 17,000$ m. The code below generates this plot, which is shown in Fig. 3.3.

```
(%i) yvt:makelist([data1[i][1],data1[i][2]],i,1,
       length(data1))$
     load(interpol)$ cspline(yvt)$ yc(x):="%$
     wxdraw2d(explicit(yc(t),t,0,tf),
       xlabel="time (s)", ylabel="height (m)")$
```

---

[3] We are using the same value for γ that we used earlier for a spherical projectile. This is a reasonable choice if we assume that our rocket has a cone-shaped nose, since the drag coefficient for a cone is very similar to that for a sphere.

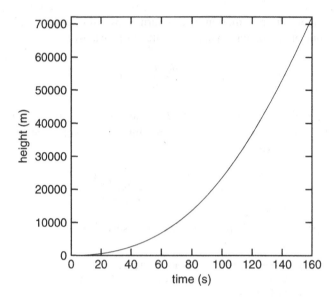

Fig. 3.3 Height of rocket as a function of time, assuming universal gravitation and quadratic air resistance

The rocket passes 17 km somewhere between 60 and 100 s after launch. To get a more precise value we can use find_root.

```
(%i)    t1:find_root(yc(t)=17000,t,60,100);
(%o)    87.592
```

The rocket reaches the top of the troposphere about 87.6 s after launch. Next we need to find the rocket's velocity at this time, which we do by fitting the velocity versus time data from rk and then evaluating the fit function at $t = 87.6$ s. We also determine the rocket's mass at this time.

```
(%i)    vvt:makelist([data1[i][1],data1[i][3]],i,1,
        length(data1))$
        cspline(vvt)$ vc(x):=''%$ vc(t1);
(%o)    476.15
```

```
(%i)    m0:2.8e6$ tf:160$ mf:6e5$ k:(m0-mf)/tf$
        m1:m0-k*t1;     (%o)   1595612.
```

The rocket's speed is 476 m/s when it leaves the troposphere, and its mass is just under 1.6 million kg. We can continue our solution from this point, but without air resistance. Again we use rk, and the last element in the output from rk provides the velocity and height of the rocket at the end of the burn.

```
(%i)    y0:17000$ v0:476.1540117582936$ tf:160-t1$ data1:
        rk([v,-GM/(R+y)^2 + k*vex/(m1-k*t)],[y,v],
        [y0,v0],[t,0,tf,0.1])$   data1[length(data1)];
(%o)    [72.4, 100292., 2219.9]
```

The rocket is just over 100 km above the Earth and traveling at about 2200 m/s. This is somewhat lower and slower than we found from our model with constant gravity and no air resistance, but it is not dramatically different. Air resistance does slow the rocket, but the effect is not very large and it is partially offset by our improved model for gravity. Even this model is far from perfect. For example, we have assumed a constant density of air throughout the troposphere when in fact the density of air decreases with altitude.

3.3 Center of Mass

So far we have addressed the motion of particles, or objects that can be treated as particles. To examine the motion of extended objects or systems of particles we must define some new concepts. One of the most important concepts for understanding the motion of an extended object or system of particles is the *center of mass*. The motion of a system can be broken down into the motion of the center of mass, and the motion of the system about the center of mass. The motion of the center of mass is determined by the external forces on the system. It is not affected by internal forces. It is for this reason that we can treat extended objects as though they were point masses: we are really examining the motion of the object's center of mass.

But what is the center of mass? How do we find it? Consider a system of N point particles with mass $m_1, m_2, \ldots, m_N$ located at positions $\vec{r}_1, \vec{r}_2, \ldots, \vec{r}_N$. The center of mass of this system is

$$\vec{R} = \frac{1}{M} \sum_{\alpha=1}^{N} m_\alpha \vec{r}_\alpha, \tag{3.20}$$

where M is the total mass of all of the particles.

Consider a system of three point particles with mass of 5, 3, and 1 kg located at $(3, -2, 7)$, $(-1, 4, 4)$, and $(6, 4, -5)$ meters, respectively. We can find the location of the center of mass for this system using Eq. 3.20.

```
(%i) m1:5$ r1:[3,-2,7]$ m2:3$ r2:[-1,4,4]$ m3:1$
     r3:[6,4,-5]$
     R:(m1*r1+m2*r2+m3*r3)/(m1+m2+m3);
(%o) [2, 2/3, 14/3]
```

The center of mass is at $(2, 2/3, 14/3)$ meters. We plot the location of the center of mass, as well as the locations of the three particles. An interactive plot is best so that you can view the 3D plot from various angles, but here we show one particular view in Fig. 3.4.[4]

---

[4]In the workbook that accompanies this section, replace wxdraw with draw. The result will be an interactive plot.

Fig. 3.4 Three point masses
(*filled circles*) and their center
of mass (*open circle*)

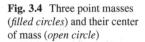

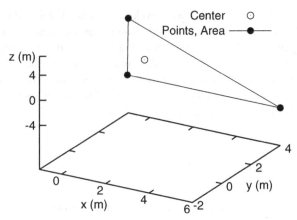

```
(%i) wxdraw3d(key="Center",point_type=circle,
     points([R]), point_type=7, points_joined=true,
     key="Points, Area", points([r1,r2,r3,r1]),
     view=[70,30], xlabel="x (m)", ylabel="y (m)",
     zlabel="z (m)", dimensions=[480,480], xtics=2,
     ytics=2, ztics=4)$
```

The three point masses form a triangle. By viewing this plot from different angles
in the interactive view you should be able to see that the center of mass lies in the
plane of this triangle, and in the triangle's interior. It is not in the center of the
triangle (however you define that center), but is instead shifted toward the locations
of the larger masses.

What about the center of mass for an extended object? Think of a solid object as
consisting of an infinite number of infinitesimally small point masses. In this case
the sum in Eq. 3.20 turns into an integral over the volume of the object:

$$\vec{R} = \frac{1}{M} \int_V \rho\, \vec{r}\, dV, \tag{3.21}$$

where dV is an infinitesimal volume element within the object, $\vec{r}$ is the location of
that element, and ρ is the density of the object (which may depend on location).
Although the term "volume" implies a three-dimensional object, this formula can
be easily extended to 2D objects by integrating over the object's area and using a
surface density. Likewise, we can write a 1D version in which we integrate over the
object's length and use a linear density.

As an example, consider a uniform density triangular lamina (flat surface) with
vertices at $(0,0)$, $(a,0)$, and $(0,b)$. Think of this triangle as the area between the
x-axis, the y-axis, and the line $y = b - bx/a$. First we need to find the mass of this
triangle. We can easily find this mass by multiplying the density ρ by the triangle's
area $A = (1/2)ab$. However, it is instructive to calculate the mass by integrating the
density over the area of the triangle.

```
(%i) integrate(integrate(%rho,y,0,b-b*x/a),x,0,a);
```
(%o) $\frac{\rho a b}{2}$

The result is exactly what we expected. Now we can calculate the x-coordinate and y-coordinate of the center of mass by evaluating the same integral except with a factor of x or y added to the integrand, and then dividing by the total mass.

```
(%i) %rho:2*M/(a*b)$ integrate(
       integrate(%rho*x,y,0,b-b*x/a),x,0,a)/M;
```
(%o) $\frac{a}{3}$
```
(%i) integrate(integrate(
       %rho*y,y,0,b-b*x/a),x,0,a)/M;
```
(%o) $\frac{b}{3}$

The center of mass of the triangle lies at $(a/3, b/3)$. We can use *Maxima* to display the triangle and its center of mass for $a = 4$ and $b = 3$, in order to illustrate the reasonableness of our result. The resulting plot is shown in Fig. 3.5.

```
(%i) wxdraw2d(fill_color=gray, polygon([[0,0],[4,0],
       [0,3]]), point_type=7,points([[4/3,1]]));
```

Now consider a three-dimensional solid. The solid is a uniform quarter sphere of mass M constructed by building a sphere of radius R centered at the origin but keeping only the portion with $y \geq 0$ and $z \geq 0$. We calculate the center of mass coordinates using Eq. 3.21, but it will be most convenient to use spherical polar coordinates. The transformation from polar to Cartesian coordinates is given by $x = r \sin\theta \cos\phi$, $y = r \sin\theta \sin\phi$, and $z = r \cos\theta$ where θ is the polar angle and ϕ is the azimuthal angle. The volume element is $dV = r^2 \sin\theta \, dr \, d\theta \, d\phi$.

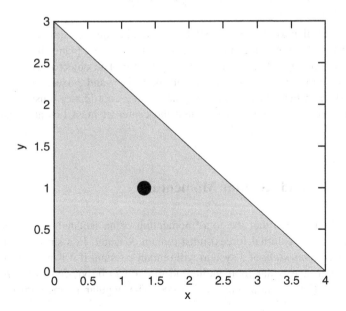

Fig. 3.5 A uniform triangular lamina (*shaded region*) and its center of mass (*filled circle*)

We want to integrate from the origin out to the surface of the sphere ($0 < r < R$), but only in the region where $z > 0$ (so $0 \leq \theta \leq \pi/2$) and $y > 0$ (so $0 \leq \phi \leq \pi$). We can integrate to find the mass of this quarter sphere:

$$M = \int_0^R \int_0^{\pi/2} \int_0^{\pi} \rho \, r^2 \sin \theta \, d\phi \, d\theta \, dr. \tag{3.22}$$

We can evaluate this triple integral in *Maxima*.

```
(%i) kill(%rho, R)$ integrate(integrate(
     integrate(%rho*r^2*sin(theta),r,0,R),
       theta,0,%pi/2),phi,0,%pi);
```
$$(\%o) \quad \frac{\pi \rho R^3}{3}$$

The mass is $M = \rho \pi R^3/3$, as expected since the volume of a quarter sphere is $\pi R^3/3$. So the density is $\rho = 3M/(\pi R^3)$. Now we can determine the coordinates for the center of mass by computing the same integral, but with an additional factor of x, y, or z (converted into spherical polar coordinates) in the integrand, and then dividing by the mass.

```
(%i) %rho:3*M/(%pi*R^3)$ X:integrate(integrate(
     integrate( %rho*r^3*sin(theta)^2*cos(phi),r,0,R),
       theta,0,%pi/2),phi,0,%pi)/M$
     Y:integrate(integrate(integrate(
       %rho*r^3*sin(theta)^2*sin(phi),r,0,R),
       theta,0,%pi/2),phi,0,%pi)/M$
     Z:integrate(integrate(integrate(
       %rho*r^3*sin(theta)*cos(theta),r,0,R),
       theta,0,%pi/2),phi,0,%pi)/M$ [X,Y,Z];
```
$$(\%o) \quad [0, \tfrac{3R}{8}, \tfrac{3R}{8}]$$

The center of mass lies at $(0, 3R/8, 3R/8)$. This quarter sphere is symmetric about the y-z plane, so its center of mass must lie in that plane. It is no surprise that the x-coordinate of the center of mass is zero. The object is also symmetric about a plane that makes a 45° angle with both the y and z axes, so the center of mass must lie in that plane. Therefore, the y and z coordinates must be the same. More precisely, our result shows us that the center of mass lies at a distance of $3R\sqrt{2}/8 \approx 0.53R$ from the origin.

3.4 Torque and Angular Momentum

We have already seen that the total momentum of an isolated system (a system not subject to any external forces) must remain constant. In a similar manner, the total *angular momentum* of a system will remain constant if it is not subject to any external *torques*. Angular momentum and torque can be viewed as the rotational analogues of (linear) momentum and force. The angular momentum of a point particle about the origin is defined as the cross product of the particle's position and momentum vectors:

$$\vec{\ell} = \vec{r} \times \vec{p}. \tag{3.23}$$

For a system of N particles, the total angular momentum of the system is

$$\vec{L} = \sum_{\alpha=1}^{N} \vec{r}_\alpha \times \vec{p}_\alpha = \sum_{\alpha=1}^{N} m_\alpha \vec{r}_\alpha \times \vec{v}_\alpha. \tag{3.24}$$

The code below shows how to use *Maxima* to compute the angular momentum of three point masses, as well as the total angular momentum of the system (with values given in MKS units). Recall that to perform cross products you must first load the `vect` package. The symbol for a cross product is a tilde, but to get *Maxima* to display the result of the cross product you must use the `express` command.

```
(%i) load(vect)$ m1:5$ r1:[3,-2,7]$ v1:[2,0,3]$
     m2:3$ r2:[-1,4,4]$ v2:[-3,4,1]$
     m3:1$ r3:[6,4,-5]$ v3:[0,-1,-5]$
     L1:m1*express(r1~v1)$
     L2:m2*express(r2~v2)$ L3:m3*express(r3~v3)$
     Lnet:L1+L2+L3$ [L1,L2,L3,Lnet];
(%o) [[-30, 25, 20], [-36, -33, 24], [-25, 30, -6], [-91, 22, 38]]
```

If we take the time derivative of Eq. 3.23 we find

$$\frac{d\vec{\ell}}{dt} = \frac{d\vec{r}}{dt} \times \vec{p} + \vec{r} \times \frac{d\vec{p}}{dt}. \tag{3.25}$$

But $d\vec{r}/dt = \vec{v}$ which is parallel to $\vec{p}$ so $\vec{v} \times \vec{p} = 0$. Also, Newton's Second Law tells us that $d\vec{p}/dt = \vec{F}$ where $\vec{F}$ is the net force on the particle. So

$$\frac{d\vec{\ell}}{dt} = \vec{\Gamma}, \tag{3.26}$$

where $\vec{\Gamma} = \vec{r} \times \vec{F}$ is the torque on the particle about the origin.

For a system of particles we have

$$\frac{d\vec{L}}{dt} = \sum_{\alpha=1}^{N} \vec{\Gamma}_\alpha = \sum_{\alpha=1}^{N} \vec{r}_\alpha \times \vec{F}_\alpha. \tag{3.27}$$

Newton's Third Law can be used to show that internal torques between particles in the system cancel each other out, and thus we need only consider external torques (torques caused by external forces) when computing the net torque on a system. For example, we can compute the net gravitational torque about the origin on the system of particles we examined above. Each particle is subject to a gravitational force of magnitude $m_\alpha \vec{g}$, where $\vec{g} = -g\hat{z}$.

```
(%i) m1:5$ r1:[3,-2,7]$ m2:3$ r2:[-1,4,4]$ m3:1$
     r3:[6,4,-5]$ grav:[0,0,g]$
     t1:m1*express(r1~grav)$
     t2:m2*express(r2~grav)$
     t3:m3*express(r3~grav)$
     tnet:t1+t2+t3$ [t1,t2,t3,tnet];
(%o) [[-10g,-15g,0],[12g,3g,0],[4g,-6g,0],[6g,-18g,0]]
```

We see that gravity produces a nonzero net torque and thus the angular momentum of this system about the origin will change. But what if we choose a different origin? For example, we could compute the torque about the center of mass of the system. If the center of mass position is $\vec{R}$ then the torque about the center of mass will be $\vec{\Gamma} = (\vec{r} - \vec{R}) \times \vec{F}$. Using our previous formula for computing the center of mass of a system of particles (Eq. 3.20) we can calculate the net torque on our system.

```
(%i) R:(m1*r1+m2*r2+m3*r3)/(m1+m2+m3)$
     t1c:m1*express((r1-R)~grav)$
     t2c:m2*express((r2-R)~grav)$
     t3c:m3*express((r3-R)~grav)$
     tnetc:t1c+t2c+t3c$       [R,t1c,t2c,t3c,tnetc];
(%o) [[2, 2/3, 14/3],[-40g/3,-5g,0],[10g,9g,0],[10g/3,-4g,0],[0,0,0]]
```

Our results indicate that the torque about the center of mass is zero. It is a special property of the center of mass that the torque on an object (or system of particles) about its center of mass by a uniform gravitational force is always zero. For this reason, the center of mass is sometimes referred to as the "center of gravity." To balance an object (and thus keep its angular momentum constant) you must support the object beneath its center of mass.

3.5 Products and Moments of Inertia

In this section we examine the angular momentum of an object or system of particles that is rotating about a fixed axis. We start by determining the angular momentum of a single particle rotating about the z-axis. The angular velocity of the rotation is given by the vector $\vec{\omega} = \omega\hat{z}$, where $|\omega|$ is the angular speed of the rotation. The right-hand rule determines the direction of rotation about that axis: point the thumb of your right hand in the direction of $\vec{\omega}$ and your fingers curl in the direction of rotation. So as viewed from above the x–y plane the rotation will be counterclockwise if $\omega > 0$, clockwise if $\omega < 0$.

The velocity of the particle that results from its rotation about a fixed axis through the origin is

$$\vec{v}_r = \vec{\omega} \times \vec{r}. \tag{3.28}$$

We can use *Maxima* to determine the velocity of our particle rotating about the z-axis.

```
(%i) load(vect)$
     w: [0,0,omega]$    r: [x,y,z]$    vr:express(w~r);
(%o) [-ω y, ω x, 0]
```

We see that the particle's velocity is directed along the x–y plane, as we should expect for a rotation about the z-axis. Likewise, we see that the speed of the particle is $\omega(x^2 + y^2) = r\omega$ as expected from the rules of circular motion. Now we can compute the angular momentum of this particle, $\vec{\ell}_r = m\vec{r} \times \vec{v}_r$.

```
(%i) Lr:m*express(r~vr);
(%o) [-m ω x z, -m ω y z, m (ω y² + ω x²)]
```

Note that the components of this angular momentum vector have a common factor of ω. The remaining factors depend on the mass and position of the particle. We can extend this result to find the components of the total angular momentum for a system of N particles:

$$\vec{L} = \omega \left(I_{xz}\hat{x} + I_{yz}\hat{y} + I_{zz}\hat{z} \right), \tag{3.29}$$

where

$$I_{xz} = -\sum_{\alpha=1}^{N} m_\alpha x_\alpha z_\alpha,$$

$$I_{yz} = -\sum_{\alpha=1}^{N} m_\alpha y_\alpha z_\alpha,$$

$$I_{zz} = \sum_{\alpha=1}^{N} m_\alpha (x_\alpha^2 + y_\alpha^2). \tag{3.30}$$

I_{zz} is known as the *moment of inertia* of the system about the z-axis, while I_{xz} and I_{yz} are known as *products of inertia*.

To show that Eq. 3.29 works we compute the angular momentum of a system of three particles (with the same positions as in the previous section) rotating about the z-axis. First we compute the total angular momentum of the system using Eqs. 3.28 and 3.23.

```
(%i) v1r:express(w~r1)$ L1r:m1*express(r1~v1r)$
     v2r:express(w~r2)$ L2r:m2*express(r2~v2r)$
     v3r:express(w~r3)$ L3r:m3*express(r3~v3r)$
     Lnetr:L1r + L2r + L3r;
(%o) [-63 ω, 42 ω, 168 ω]
```

Now we compute the total angular momentum of the rotating system of particles using Eq. 3.29 to show that it gives the same result.

```
(%i) Ixz:-m1*r1[1]*r1[3]-m2*r2[1]*r2[3]-
        m3*r3[1]*r3[3]$
     Iyz:-m1*r1[2]*r1[3]-m2*r2[2]*r2[3]-m3*r3[2]*r3[3]$
     Izz:m1*(r1[1]^2+r1[2]^2)+m2*(r2[1]^2+r2[2]^2)+
        m3*(r3[1]^2+r3[2]^2)$
     Lnetr:omega*[Ixz,Iyz,Izz];
(%o) [-63ω, 42ω, 168ω]
```

We can extend the definitions of the moment and products of inertia to cover solid objects by converting sums into integrals.

$$I_{xz} = -\int_V \rho\, x\, z\, dV,$$

$$I_{yz} = -\int_V \rho\, y\, z\, dV,$$

$$I_{zz} = \int_V \rho\, (x^2 + y^2) dV. \tag{3.31}$$

We now apply these formulas to determine the angular momentum of the uniform solid quarter sphere that we examined in Sect. 3.3, rotating about the z-axis with angular velocity ω. We will use spherical polar coordinates to evaluate the integrals, just as we did to find the center of mass in Sect. 3.3. The density ρ and the limits of integration are the same as the center of mass integrals, only the integrand is different.

```
(%i) kill(all)$ %rho:3*M/(%pi*R^3)$
     Ixz:-integrate(integrate(integrate( %rho*r^4*
        sin(theta)^2*cos(theta)*cos(phi),r,0,R),
        theta,0,%pi/2),phi,0,%pi)$
     Iyz:-integrate(integrate(integrate( %rho*r^4*
        sin(theta)^2*cos(theta)*sin(phi),r,0,R),
        theta,0,%pi/2),phi,0,%pi)$
     Izz:integrate(integrate(integrate(%rho*r^4*
        sin(theta)^4,r,0,R),theta,0,%pi/2),phi,0,%pi)$
     L:omega*[Ixz, Iyz, Izz];
(%o) [0, -\frac{2\omega M R^2}{5\pi}, \frac{9\pi\omega M R^2}{80}]
```

Note that the angular momentum of this rotating object *does not* point in the same direction as its angular velocity. Mathematically this result follows from the fact that the object has nonzero products of inertia (I_{xz} is zero, but I_{yz} is not). It is not hard to see that as the object rotates about the z-axis the products of inertia will change, because the object's orientation relative to the x and y axes will change. Therefore, the object's angular momentum must change as it rotates. So a continual external torque must be applied for this quarter sphere to rotate at a constant angular velocity about the z-axis. In contrast, both products of inertia will be zero for an object that is symmetric about the z-axis (such as a full sphere, or the half of the sphere with $z > 0$). If such a symmetric object rotates about the z-axis then its angular momentum will point along the z-axis and the object will be able to rotate without changing its angular momentum. In that case there is no external torque needed to maintain the rotation.

3.6   Work and Potential Energy

In this section we begin working toward the Principle of Conservation of Energy. We begin by defining the kinetic energy of a particle of mass m moving with speed v to be $K = (1/2)mv^2$. If the particle is subject to a net force $\vec{F}$ and the particle is displaced by $d\vec{r}$, then Newton's Second Law can be used to show that the particle's kinetic energy will change by

$$dK = \vec{F} \cdot d\vec{r}. \tag{3.32}$$

The right-hand side of this equation is called the *work* done on the particle by the net force $\vec{F}$ over the displacement $d\vec{r}$. Therefore, Eq. 3.32 says that the change in a particle's kinetic energy is equal to the work done on the particle by the net force, a statement that is sometimes called the "Work-KE Theorem."

If a particle moves through a force field so that at each location $\vec{r}$ it is subject to a force $\vec{F}(\vec{r})$ then the total change in the particle's kinetic energy as it moves along a path from point $\vec{r}_0$ to point $\vec{r}$ is

$$\Delta K = K_{\vec{r}} - K_{\vec{r}_0} = \int_{\vec{r}_0}^{\vec{r}} \vec{F}(\vec{r}') \cdot d\vec{r}'. \tag{3.33}$$

where the integral is a line integral that is evaluated along the particle's path from $\vec{r}_0$ to $\vec{r}$.

As an example, consider a particle moving through a force field given by

$$\vec{F}_1 = 3kxy^3\hat{x} + 3kx^2y^2\hat{y}. \tag{3.34}$$

How much work is done on the particle if it moves from the origin to the point (a, b) along the path P_1 which consists of two straight line segments: one from $(0, 0)$ to $(a, 0)$ and another from $(a, 0)$ to (a, b). On the first segment the motion is entirely in the x-direction, so $\vec{F}_1 \cdot d\vec{r} = F_{1x}dx$. But on the first part of this path $y = 0$, so $F_{1x} = 0$. On the second part of the path the motion is in the y-direction, so $\vec{F}_1 \cdot d\vec{r} = F_{1y}dy$. On this part of the path $x = a$ so $F_{1y} = 3ka^2y^2$. Computing the integral along each portion of the path and adding the results show the work done along the full path.

```
(%i)  F1x(x,y):=3*k*x*y^3$ F1y(x,y):=3*k*x^2*y^2$
      integrate(F1x(x,0),x,0,a)+
      integrate(F1y(a,y),y,0,b);        (%o)  a² b³ k
```

Suppose instead that the particle moves along the path P_2, which follows the curve $y = bx^3/a^3$ from $(0,0)$ to (a, b). *Maxima*'s `diff` command can evaluate the vector $d\vec{r}$ along this path.

```
(%i) declare([a,b],constant)$ y2:b*x^3/a^3$
     dr:diff([x,y2]);          (%o) [del(x), 3bx² del(x)/a³]
```

Note the use of the `declare` command to indicate that a and b are constants, not variables. The expression `del(x)` just represents the differential element for x, or dx. Now we can evaluate $\vec{F}_1 \cdot d\vec{r}$ by substituting $y = bx^3/a^3$ into our force field equation and carrying out the dot product.

```
(%i) F1:[F1x(x,y2),F1y(x,y2)]$        dT1:F1.dr;
(%o) 12b³ kx¹⁰ del(x)/a⁹
```

Now we compute the total work along the path P_2 by integrating our result for $\vec{F}_1 \cdot d\vec{r}$ with respect to x from 0 to a.

```
(%i) integrate(12*b^3*k*x^10/a^9,x,0,a);    (%o) 12a²b³k/11
```

The work done on the particle by this force field is different along path P_2 than along path P_1. Let's try this again with a slightly different force field:

$$\vec{F}_2 = 2kxy^3\hat{x} + 3kx^2y^2\hat{y}. \tag{3.35}$$

Now we can evaluate the work done on the particle by this force field along the two paths, P_1 and P_2.

```
(%i) F2x(x,y):=2*k*x*y^3$ F2y(x,y):=3*k*x^2*y^2$
     integrate(F2x(x,0),x,0,a)+
     integrate(F2y(a,y),y,0,b);           (%o) a² b³ k
(%i) F2:[F2x(x,y2),F2y(x,y2)]$           F2.dr;
(%o) 11b³ kx¹⁰ del(x)/a⁹

(%i) integrate(11*b^3*k*x^10/a^9,x,0,a);  (%o) a² b³ k
```

In this case the work is the same along the two paths. In fact, for this force field the work will be the same along *any* path with the same starting and ending points. In such a case we say that the work is independent of the path, and a force field that has this property is called a *conservative* force field. Although we not prove it here, it turns out that a force field is conservative if and only if it has zero curl: $\vec{\nabla} \times \vec{F} = 0$. We can illustrate this property for the two force fields considered above. We can use *Maxima* to evaluate the curl with the `vect` package. Keep in mind that the curl is defined only for three-dimensional vectors, so we need to add a zero z-component to our 2D force fields in order to evaluate the curl.

```
(%i) load(vect)$   F1:[F1x(x,y),F1y(x,y),0]$  curl(F1);
(%o) curl([3kxy³, 3kx²y², 0])
```

The output from `curl` is not particularly helpful. As with other commands in the `vect` package, if we want *Maxima* to show the result of evaluating the curl we must use the `express` command.

(%i) **express(curl(F1));**
(%o) $[-\frac{d}{dz}(3kx^2y^2), \frac{d}{dz}(3kxy^3), \frac{d}{dx}(3kx^2y^2) - \frac{d}{dy}(3kxy^3)]$

Even this result is not completely satisfactory. We would like for *Maxima* to evaluate the derivatives. To do so we need to use the `ev` command and instruct *Maxima* to evaluate anything that involves the `diff` operator. The code for this evaluation is shown below.

(%i) **ev(express(curl(F1)),diff);**   (%o) $[0, 0, -3kxy^2]$

So $\vec{\nabla} \times \vec{F}_1 = -3kxy^2\hat{z} \neq 0$. Therefore $\vec{F}_1$ is not conservative, and the work done by this force is not path-independent as we showed earlier. Now we can evaluate $\vec{\nabla} \times \vec{F}_2$.

(%i) **F2:[F2x(x,y),F2y(x,y),0]$**
        ev(express(curl(F2)),diff);   (%o) $[0, 0, 0]$

Here we see that $\vec{\nabla} \times \vec{F}_2 = 0$. Therefore, this force field is conservative and the work done by $\vec{F}_2$ will be path-independent, in agreement with our earlier results. When the work is path-independent, then the kinetic energy of a particle moving from $\vec{r}_0$ to $\vec{r}$ will change by the same amount regardless of the path taken. In that case it is useful to define a *potential energy* associated with the conservative force field.

The potential energy function $U(\vec{r})$ is defined such that

$$U(\vec{r}) - U(\vec{r}_0) = -\int_{\vec{r}_0}^{\vec{r}} \vec{F}(\vec{r}') \cdot d\vec{r}', \tag{3.36}$$

where $\vec{r}_0$ is an arbitrary point at which potential energy will be zero. The right side of this equation is just the negative of the work done on the particle as it moves from $\vec{r}_0$ to $\vec{r}$. Thus, the change in the potential energy is the negative of the change in kinetic energy, and therefore the *total mechanical energy* $E = U + K$ will be conserved.

We have already calculated that the work done by $\vec{F}_2$ from the origin to (a, b) is ka^2b^3. Therefore, if we define the origin as our zero point for potential energy the potential energy function associated with $\vec{F}_2$ must be $U_2(x, y) = -kx^2y^3 + C$, where C is a constant. We can verify that the change in potential energy as the particle moves from $(0, 0)$ to (a, b) is just the negative of the work we calculated earlier.

(%i) **U(x,y):=-k*x^2*y^3+C$**   **U(a,b)-U(0,0);**
(%o) $-a^2b^3k$

We can even recover the force field from the potential energy function by using the inverse of Eq. 3.36:

$$\vec{F} = -\vec{\nabla}U(\vec{r}). \tag{3.37}$$

The code below computes the negative gradient of $U(x, y)$ to show that this gives the correct force field $\vec{F}_2(x, y)$.

```
(%i) ev(express(-grad(U(x,y)))),diff);
(%o) [2kxy³, 3kx²y², 0]
```

3.7   Fall from a Great Height: Conservation of Energy

In this section we use the Principle of Conservation of Energy in order to examine the behavior of an object falling to Earth from a great height. We will ignore the effects of Earth's atmosphere, so we do not account for air resistance. Air resistance is a velocity-dependent force that is nonconservative, so including air resistance would prevent us from making use of the conservation of energy. To examine the fall of an object with air resistance we could numerically solve Newton's Second Law as was done in Sect. 2.3.

To use the conservation of energy, we must first make sure that we are dealing with conservative forces. In studying the fall of an object we will consider two different models for the gravitational force. Our first model examines a uniform gravitational force given by $\vec{F}_g = -mg\hat{z}$ where $\hat{z}$ is a unit vector pointing upward (perpendicular to Earth's surface), m is the mass of the object, and $g \approx 9.8$ m/s^2 is the gravitational field strength at Earth's surface. We can make sure that this force is conservative by evaluating the curl.

```
(%i) load(vect) $        Fg: [0,0,-m*g] $
     ev(express(curl(Fg)),diff);               (%o)  [0,0,0]
```

Since $\vec{\nabla} \times \vec{F}_g = 0$ we know that the force is conservative. Now we can find the potential energy function associated with this force. We choose the Earth's surface ($z = 0$) to be the zero-point for this potential energy function, and we integrate along a path that goes straight up from Earth's surface so that $\vec{F}_g \cdot d\vec{r} = (-mg)dz$ and

$$U(z) = -\int_0^z (-mg)dz' = mgz. \qquad (3.38)$$

The object is subject only to the conservative gravitational force, so the total mechanical energy, $E = K + U$, is conserved.

If the object is dropped from rest at a height h above Earth's surface, then $E = mgh$. We can determine the speed v of the object at any other height z by solving the equation $mgh = (1/2)mv^2 + mgz$.

```
(%i) spd:solve(m*g*h=m*g*z+(1/2)*m*v^2,v);
(%o) [v = -√2 √(gh-gz), v = √2 √(gh-gz)]
```

There are two solutions, but they differ only by sign. The positive solution tells us the speed, and we know that the object will move in the $-\hat{z}$ direction. Therefore, the velocity of the object at height z is

$$v = -\sqrt{2g(h-z)}\hat{z}. \tag{3.39}$$

We use this result to determine the speed on impact with Earth if the object was dropped from the height of the International Space Station,[5] which is about 370 km.

```
(%i)  h:370000$        g:9.8$        sqrt(2*g*h);
(%o)  2693.0
```

This object would hit the Earth at a speed of almost 2.7 km/s. We examine how the speed of the object changes as it falls by constructing a plot of speed versus height using the code below. The resulting plot is shown in Fig. 3.6.

```
(%i)  wxdraw2d(explicit(sqrt(2*g*(h-y)),y,0,h),
          xlabel="height (m)", ylabel="speed (m/s)",
          xtics=100000)$
```

Figure 3.6 illustrates that the increase in speed is much greater in the first kilometer of fall than in the last kilometer. This makes sense because with constant

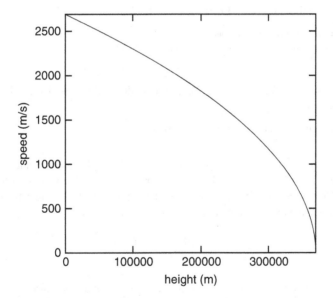

Fig. 3.6 Speed as a function of height for an object dropped from the height of the ISS, assuming constant gravity and no air resistance

[5]The ISS is in orbit around the Earth, so it is *not* at rest. We are assuming the object is dropped from rest, so our calculation does not apply to an object that is actually dropped off of the ISS. In fact, such an object would continue to orbit Earth right along with the ISS.

acceleration the speed increases at a constant rate with respect to time. At the beginning of its fall the object is moving slower and therefore it takes a longer time to cover the first kilometer. During this long time the object will significantly increase its speed. Just before hitting Earth the object is moving very fast, so it covers the last kilometer very quickly and does not gain much speed during this brief interval of time.

To find the total time of fall for this object we rewrite Eq. 3.39 as

$$\frac{dz}{dt} = -\sqrt{2g(h-z)}. \tag{3.40}$$

Separating the variables and integrating both sides we have

$$\int_0^T dt = -\int_h^0 \frac{dz}{\sqrt{2g(h-z)}} \tag{3.41}$$

or

$$T = \frac{1}{\sqrt{2g}} \int_0^h \frac{dz}{\sqrt{h-z}}. \tag{3.42}$$

We evaluate the integral in Eq. 3.42 with *Maxima*.

```
(%i) kill(values)$ assume(h>0)$
     integrate(1/sqrt(h-y),y,0,h)/sqrt(2*g);
(%o)  √2 √h
      ─────
       √g
```

The total time of fall is $T = \sqrt{2h/g}$. We evaluate this time for a fall from the height of the ISS.

```
(%i) h:370000$ g:9.8$ sqrt(2*h/g);
(%o) 274.79
```

Our object takes almost 275 s to fall from the ISS height to the surface of Earth.

Of course, Earth's gravitational field is not truly uniform. It changes with distance from Earth's center. A more accurate model for the fall of an object from a great height would use Newton's Universal Law of Gravitation:

$$\vec{F}_N = \frac{-GMm}{r^2}\hat{r} = \frac{-GMm}{r^3}\vec{r}, \tag{3.43}$$

where G is Newton's gravitational constant, M is the mass of Earth, and $\vec{r}$ is the position of the object measured relative to Earth's center. We can show that this force is conservative by showing that it has zero curl.

```
(%i) FN:-GM*m/(x^2+y^2+z^2)^(3/2)*[x,y,z]$
     ev(express(curl(FN)),diff);
(%o) [0,0,0]
```

If we set the zero point of potential energy at the surface of Earth ($z = R$, where R is Earth's radius) then our potential energy function is

$$U(z) = -\int_{\vec{r_0}}^{\vec{r}} \vec{F}_N \cdot d\vec{r} = -\int_{R}^{R+z} \frac{-GMm}{z'^2} dz'. \tag{3.44}$$

We use *Maxima* to evaluate this integral.

```
(%i) assume(z>0,R>0)$ -integrate(-GM*m/zp^2,zp,R,R+z);
```
$(\%o)$ m GM $\left(\frac{1}{R} - \frac{1}{R+z}\right)$

We now have

$$U(z) = GMm\left(\frac{1}{R} - \frac{1}{R+z}\right), \tag{3.45}$$

so the total energy of an object dropped from rest at height h is $E = GMm(1/R - 1/(R + h))$ and we can solve for the speed as a function of height by solving $E = K + U$ for the speed v.

```
(%i) kill(h)$ solve(GM*m*(1/R-1/(R+h))=
      GM*m*(1/R-1/(R+z))+(1/2)*m*v^2,v);
```
$(\%o)$ $[v = -\frac{\sqrt{2}\sqrt{h\,GM-z\,GM}}{\sqrt{R^2+zR+hR+hz}}, \quad v = \frac{\sqrt{2}\sqrt{h\,GM-z\,GM}}{\sqrt{R^2+zR+hR+hz}}]$

This speed formula is much more complicated than our result for a uniform gravitational field. We can use this formula to evaluate the speed on impact with Earth's surface ($z = 0$).

```
(%i) R:6371000$ h:370000$ GM:3.983324e14$
      sqrt(2*h*GM/(R^2+h*R));
```
$(\%o)$ 2619.8

In this model the impact speed is just over 2.6 km/s, somewhat slower than our result for a uniform field. This makes sense because the universal gravitation model has a weaker gravitational force, and thus a smaller acceleration, everywhere but at Earth's surface. We can construct a plot of speed versus time for this model as well, using the code below. The resulting plot is shown in Fig. 3.7.

```
(%i) wxdraw2d(explicit(sqrt(2)* sqrt((h*GM)/
      (R^2+y*R+h*R+h*y))-(y*GM)/(R^2+y*R+h*R+h*y)),
      y,0,h) ,xlabel="height (m)", ylabel=
      "speed (m/s)", xtics=100000)$
```

The plot in Fig. 3.7 looks much like the plot from our uniform field model. This result is not surprising because even the height of the ISS is small relative to Earth's radius, so the difference between universal gravitation and uniform gravity is slight. Now we try to find the total time of fall for our universal gravitation model. If we use separation of variables as before we find that

$$T = \frac{1}{\sqrt{2GM}} \int_{0}^{h} \left(\sqrt{\frac{1}{R+z} - \frac{1}{R+h}}\right)^{-1} dz. \tag{3.46}$$

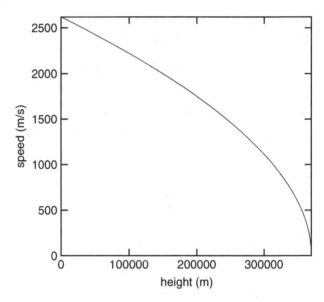

Fig. 3.7 Speed as a function of height for an object dropped from the height of the ISS, assuming universal gravitation and no air resistance

We can try to evaluate this integral using *Maxima*.

```
(%i)  integrate(1/sqrt(1/(R+y)-1/(R+h)),y,0,h)/
      sqrt(2*GM);
```
$$(\%o) \quad 3.5429 \, 10^{-8} \left(-13500 \, 7490^{3/2} \text{asin} \left(\tfrac{6001}{6741} \right) + 6750 \, 7490^{3/2} \pi + 30000 \sqrt{7490} \sqrt{2357270} \right)$$

```
(%i)  float(%);
(%o)  287.87
```

Recent versions of *Maxima* return the correct result, 287.87 s, as shown above. However, older versions of *Maxima* may be unable to evaluate this integral. In situations where *Maxima* cannot evaluate an integral analytically, we can still use *Maxima* to evaluate the integral numerically. *Maxima* includes multiple numerical integration routines. For more information on numerical integration methods, see Sect. A.2.

For our problem we can use the quad_qags routine, which uses adaptive interval subdivision to evaluate the integral of a general function over a finite interval. The arguments of quad_qags include the function to be integrated, the integration variable, and the lower and upper limits of integration, in that order.

```
(%i)  R:6371000$ h:370000$ GM:3.983324e14$
      quad_qags(1/sqrt(2*GM*(1/(R+y)-1/(R+h))),y,0,h);
(%o)  [287.87, 3.15072 10⁻⁹, 315, 0]
```

We see that the output from quad_qags consists of four numbers. The first is the approximate value for our integral, 287.87 s, in agreement with the analytical result obtained above. Note that this time is slightly longer than for our uniform gravity model, which again makes sense because universal gravitation is weaker than our uniform gravitational force above Earth's surface. The second number in the output from quad_qags is the estimated absolute error of the approximation. In this case, the absolute error of 3.2×10^{-9} s represents a relative error of about one part in 100 billion. The third number in the output shows how many times the integrand had to be evaluated and the fourth number is an error code (0 indicates no problems). For more information on quad_qags see Sect. A.2 or *Maxima*'s help menu.

3.8 Exercises

1. Two rubber balls are placed one on top of the other. The top ball has mass m and the bottom ball has mass M. The balls are dropped onto a hard floor. The bottom ball strikes the floor with speed v_0. The elastic collision between the bottom ball and the floor simply reverses the velocity of the bottom ball. The bottom ball then undergoes a head-on elastic collision with the top ball. Determine the final speeds of the two balls after this collision. If the bottom of the bottom ball was initially at a height h above the floor when the balls were dropped, how high will the top ball bounce on its rebound? How high can the top ball go in the limit where the mass of the bottom ball becomes much greater than that of the top ball?

2. Consider a glancing elastic collision between a particle of mass m_1, initially moving in the x-direction at speed v_0, and a stationary particle of mass m_2. If the first particle is deflected by an angle θ, determine the speed of the first particle after the collision. (Hint: you must also consider the x- and y-components of the second particle's velocity after the collision.)

3. In Sect. 3.2 we analyzed the first stage of a Saturn V rocket. Once the first stage is complete the spent S-IC engine is released, reducing the mass of the rocket by 1.4×10^5 kg. Then the stage 2 (S-II) engine fires for 6 min. This engine expels exhaust at 4200 m/s and contains 4.4×10^5 kg of fuel.

   (a) Determine the final speed of the rocket at the end of the stage 2 burn if the rocket was launched in deep space (with no gravitational forces). Start with the relevant stage 1 results from Sect. 3.2.

   (b) Determine the final speed of the rocket, and its height above Earth's surface, if the rocket was launched from Earth. Start with the stage 1 results from Sect. 3.2 that incorporated air resistance in the troposphere and universal gravitation. Solve using rk, using Newton's universal law of gravitation.

4. Consider a hemispherical shell (the portion of a spherical shell centered on the origin that is above the x–y plane) of mass M, inner radius a, and outer radius b. Find the center of mass of this object. Also, find the products and moment of inertia for rotation about the z-axis (I_{xz}, I_{yz}, and I_{zz}). What is the angular

momentum of this object if it rotates about the z-axis with angular velocity ω? Is an external torque required to keep this object rotating this way? Explain how you might have anticipated this result based on the symmetry of the object. Finally, go back and reevaluate your results in the limit of a thin spherical shell (i.e., the limit $a \to b$).

5. Consider the force field $\vec{F} = -kx^2y^4\hat{x} - 4kx^3y^3\hat{y}$. Evaluate the work done on a particle that moves through this field along a path that consists of two straight line segments: from $(0,0)$ to $(0,b)$ and then from $(0,b)$ to (a,b). Then find the work done if the particle takes a straight line path from $(0,0)$ to (a,b). Is the work done along the two paths the same? Evaluate the curl of this force field and explain how your result for the curl relates to whether or not the work done along the two paths is the same. Is this a conservative force field?

6. Consider the potential energy function $U(x,y) = kx^3y^4$. Find the force field associated with this potential energy. Evaluate the work done on a particle that moves through this field along a path that consists of two straight line segments: from $(0,0)$ to $(0,b)$ and then from $(0,b)$ to (a,b). Then find the work done if the particle takes a straight line path from $(0,0)$ to (a,b). Is the work done along the two paths the same? Show that the curl of this force field is zero.

7. A block of mass m is moving at speed v_0 when it hits a spring. The spring, initially in its equilibrium position, is compressed by the impact of the block. If the spring exerts a Hooke's Law force $(F = -kx)$ on the block, determine the potential energy function $U(x)$ where x is the displacement of the spring from its equilibrium position. Use conservation of energy to find the speed of the block as a function of x, and plot this function. How far will the spring be compressed before the block comes to rest? Use separation of variables to calculate how much time it takes for the block to come to rest after hitting the spring. Compare your result to the known period of a mass oscillating on a spring $(T = 2\pi\sqrt{m/k})$. Does your result makes sense?

Chapter 4
Oscillations

In this chapter we examine oscillating systems. In particular, we focus on harmonic oscillations produced by linear forces. We begin by showing why such oscillations are common and then proceed to an analysis of oscillating systems with and without damping and driving forces. The chapter concludes with a brief examination of an oscillating system with nonlinear forces, namely the simple pendulum.

4.1 Stable and Unstable Equilibrium Points

Section 3.6 shows that the force on a particle in a conservative system can be derived from the potential energy function of the system:

$$\vec{F} = -\vec{\nabla} U. \tag{4.1}$$

This relationship implies that the system has *equilibrium point* (a point at which there is no force on the particle) whenever $\vec{\nabla} U = 0$. In a one-dimensional system the criterion for an equilibrium point at $x = a$ is just

$$\left(\frac{dU(x)}{dx} \right)_{x=a} = 0. \tag{4.2}$$

In this section we examine the nature of equilibrium points in one-dimensional systems. The extension to higher dimensions is straightforward, at least in Cartesian coordinates.

To understand the different types of equilibrium points we examine the motion of the particle in the vicinity of $x = a$. As long as we remain sufficiently close to this equilibrium point, then we can accurately approximate the potential energy function for the system by using a Taylor series expansion about $x = a$ with only a

© Todd Keene Timberlake & J. Wilson Mixon, Jr. 2016
T.K. Timberlake, J.W. Mixon, *Classical Mechanics with Maxima*, Undergraduate
Lecture Notes in Physics, DOI 10.1007/978-1-4939-3207-8_4

few terms. We can use *Maxima* to compute the Taylor series expansion for a generic $U(x)$ to second order.

(%i) `taylor(U(x),x,a,2);`

(%o) $U(a) + \left(\frac{d}{dx} U(x)\big|_{x=a}\right)(x-a) + \frac{\left(\frac{d^2}{dx^2} U(x)\big|_{x=a}\right)(x-a)^2}{2} + \ldots$

Since we are free to define the zero-point for our potential energy function, we can make the first term in this Taylor series vanish by choosing $U(a) = 0$. The second term vanishes because $x = a$ is an equilibrium point, satisfying Eq. 4.2. Therefore, to second order in $(x - a)$ we have

$$U(x) \approx (1/2)k(x - a)^2, \tag{4.3}$$

where $k = (d^2 U/dx^2)_{x=a}$. We can use Eq. 4.1 to evaluate the approximate force function.

(%i) `Ua(x):=(1/2)*k*(x-a)^2$ Fa(x):="(-diff(Ua(x),x));`
(%o) $Fa(x) := -k\,(x-a)$

So $F(x) = -k(x - a)$. If $k > 0$ then this force has the form of Hooke's Law ($F = -kx$) and it serves as a restoring force that always pushes the particle back toward the equilibrium point. In this case we say that the equilibrium point is *stable*, and the motion of the particle near the equilibrium point will consist of oscillations like those of a spring. If $k < 0$ then the force will push the particle farther away from the equilibrium point and we say that such an equilibrium point is *unstable*. If $k = 0$ we cannot determine whether the equilibrium point is stable or unstable and we must consider higher-order terms in our Taylor series expansion of $U(x)$.

Now we can identify and analyze the equilibrium points of any one-dimensional system for which we know the potential energy function. For example, consider a system with potential energy function

$$U(x) = ax^3 - bx^2 + c. \tag{4.4}$$

We can find the equilibrium points by solving for the values of x such that $dU/dx = 0$.

(%i) `U(x):=a*x^3+b*x^2+c$ solve("(diff(U(x),x))=0,x);`
(%o) $[x = -\frac{2b}{3a}, x = 0]$

So the equilibrium points are at $x = -2b/(3a)$ and $x = 0$. We evaluate the stability of these equilibrium points by calculating d^2U/dx^2 at each point.

(%i) `[k(x):="(diff(U(x),x,2)),  k(0),  k(-2*b/(3*a))];`
(%o) $[k(x) := 6ax + 2b, \quad 2b, \quad -2b]$

At $x = 0$ $k = 2b$, so this equilibrium point is stable if $b > 0$ and unstable if $b < 0$. Likewise, at $x = -2b/(3a)$ $k = -2b$, so that equilibrium point is stable if $b < 0$ and unstable if $b > 0$. Figure 4.1, which shows a specific example of this potential energy function, helps us see how this works.

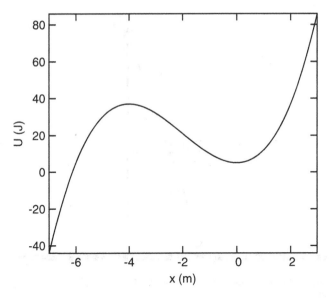

Fig. 4.1 A potential energy function with a stable equilibrium point at $x = 0$ and an unstable equilibrium point at $x = -4$

```
(%i) wxdraw2d(explicit(subst([a=1,
       b=6,c=5],U(x)), x,-7,3),xlabel="x (m)", ylabel="U (J)");
```

In this case $a = 1$ and $b = 6$, so there should be an unstable equilibrium point at $x = -4$ and a stable equilibrium point at $x = 0$. The graph shows that the potential energy function has a local maximum at $x = -4$ and a local minimum at $x = 0$. In this case, therefore, stable equilibrium points correspond to local minima in $U(x)$, while unstable equilibrium points correspond to local maxima.

Although we will not present a proof here, it turns out that this technique for finding and analyzing equilibrium points can be applied to many curvilinear systems. Curvilinear systems are systems where motion is constrained to take place along a curved, one-dimensional path which can be parameterized by a single coordinate. Consider the quarter sphere we examined in Sect. 3.3. We found that the center of mass of this quarter sphere lies a distance $d = (3\sqrt{2}/8)R$ from the center point of the full sphere, where R is the sphere's radius.

Now imagine setting this object onto a flat table with the curved surface of the quarter sphere touching the table. The object can rock from side to side. The geometry of this situation is illustrated in Fig. 4.2. In Fig. 4.2 the point C is the center of the full sphere. The point CM is the center of mass of the quarter sphere.

The gravitational potential energy of the system is given by mgh, where m is the mass of the quarter sphere and h is the height of the center of mass above the table. From trigonometry we find that the center of mass lies a distance $d \cos \theta$ below the point C, so that $h = R - d \cos \theta$. Therefore, the potential energy function for this system is

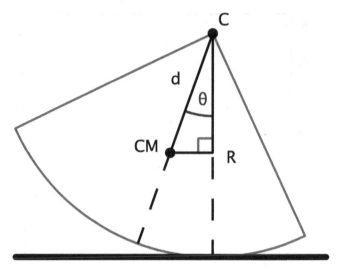

Fig. 4.2 The rocking quarter sphere. Here C is the center of the sphere and CM is the center of mass of the quarter sphere

$$U(\theta) = mg(R - d\cos\theta) = mgR(1 - 3\sqrt{2}\cos\theta/8). \qquad (4.5)$$

To find the equilibrium points we first evaluate $dU/d\theta$.

```
(%i)  U(theta):=m*g*R*(1-3*sqrt(2)*cos(theta)/8)$
      diff(U(theta),theta);
```
$$(\%o)\quad \frac{3\,g\,m\,\sin(\theta)\,R}{2^{\frac{5}{2}}}$$

The equilibrium points occur when $\sin\theta = 0$, so they occur whenever $\theta = n\pi$, where n is an integer. The only one of these angles that makes sense in this situation is $\theta = 0$, because a rotation of π radians in either direction would cause our quarter sphere to flip over onto one of its flat faces. It is not surprising that the equilibrium position occurs when the center of mass is directly below the center of rotation. But is this equilibrium stable or unstable? We evaluate $d^2U/d\theta^2$ at $\theta = 0$ to find out.

```
(%i)  k(theta):=''(diff(U(theta),theta,2))$          k(0);
```
$$(\%o)\quad \frac{3\,g\,m\,R}{2^{\frac{5}{2}}}$$

Here, $k = 3mgR/(4\sqrt{2}) > 0$, so the equilibrium point is stable. If we nudge the quarter sphere away from its equilibrium position it will just rock back and forth. It is to just such motion about a stable equilibrium point that we now turn our attention.

4.2 Simple Harmonic Motion

We have seen that the force on a particle near a stable equilibrium point at $x = a$ can be approximated as $F(x) \approx -k(x - a)$, where $k > 0$. We can define our coordinates such that $a = 0$ in order to reduce this force approximation to the familiar Hooke's Law: $F \approx -kx$. For motion near a stable equilibrium point, Newton's Second Law is approximately

$$m\ddot{x} = -kx. \qquad (4.6)$$

We use `desolve` to solve this ordinary differential equation, for a particle with initial position x_0 and initial velocity v_0.

```
(%i) assume(k>0,m>0)$          atvalue(x(t),t=0,x0)$
     atvalue('diff(x(t),t),t=0,v0)$
     eq1:m*'diff(x(t),t,2)=-k*x(t)$
     sol:desolve(eq1,x(t));
```

$$(\%o) \; x(t) = \frac{m\cos\left(\frac{\sqrt{k}t}{\sqrt{m}}\right)x0 + \frac{m^{\frac{3}{2}}\sin\left(\frac{\sqrt{k}t}{\sqrt{m}}\right)v0}{\sqrt{k}}}{m}$$

If we define $\omega_0 = \sqrt{k/m}$ then our solution reduces to

$$x(t) = \frac{v_0}{\omega_0}\sin(\omega_0 t) + x_0\cos(\omega_0 t). \qquad (4.7)$$

This solution can also be written as $x(t) = A\cos(\omega_0 t - \delta)$. To see how this works, we can use `trigexpand` to rewrite this new form of our solution.

```
(%i) trigexpand(A*cos(omega[0]*t-delta));
(%o) (sin(δ) sin(ω₀t) + cos(δ) cos(ω₀t)) A
```

From this result it is clear that the two forms of the solution will match as long as $A\sin\delta = v_0/\omega_0$ and $A\cos\delta = x_0$. Squaring these two expressions and adding them together results in $A^2 = (v_0/\omega_0)^2 + x_0^2$. Dividing the first expression by the second yields $\tan\delta = v_0/(\omega_0 x_0)$. Figure 4.3 illustrates the relationship between the parameters in the two forms of our solution.

A system with a Hooke's Law force is known as a simple harmonic oscillator, and its motion is known as simple harmonic motion. Our solution indicates that the motion is a sinusoidal oscillation with angular frequency $\omega_0 = \sqrt{k/m}$. We can visualize the motion of this system by defining and plotting functions for $x(t)$ and

Fig. 4.3 A right triangle illustrating relations between the parameters in the two forms of the simple harmonic oscillator solution

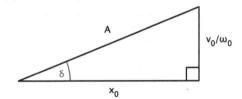

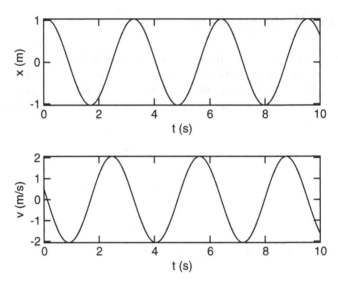

Fig. 4.4 Plots of position and velocity as a function of time for a simple harmonic oscillator

$v(t)$. The graphs in Fig. 4.4 illustrate the motion for $\omega_0 = 2$ rad/s, $x_0 = 1$ m, and $v_0 = 0.5$ m/s.

```
(%i) x(t):=(v0/omega[0])*sin(omega[0]*t)+
       x0*cos(omega[0]*t);      v(t):="(diff(x(t),t));
     xvalues:gr2d(explicit(subst([x0=1,v0=0.5,
       omega[0]=2],x(t)),t,0,10),xlabel="t (s)",
       ylabel="x (m)",ytics=1)$
     vvalues: gr2d(explicit(subst([x0=1,v0=0.5,
       omega[0]=2],v(t)), t,0,10),xlabel="t (s)",
       ylabel="v (m/s)",ytics=1)$
     wxdraw(xvalues,vvalues)$
```
$(\%o)\ \mathrm{x}(t) := \frac{v0}{\omega_0}\sin(\omega_0 t) + x0\cos(\omega_0 t)$
$(\%o)\ \mathrm{v}(t) := \cos(\omega_0 t)\,v0 - \omega_0\sin(\omega_0 t)\,x0$

It is useful to visualize this motion in *phase space*. For a one-dimensional system, phase space is a two-dimensional space with position on one axis and velocity on another. The code below shows how to generate a plot of the phase space trajectory of our simple harmonic oscillator using a parametric plot. Figure 4.5 shows the phase space trajectory for our example oscillator.

```
(%i) [xt_expression,vt_expression]:
       subst([x0=1,v0=0.5,omega[0]=2], [x(t),v(t)])$
     wxdraw2d(nticks=100,parametric(xt_expression,
       vt_expression,t,0,%pi), xlabel="x (m)",
       ylabel="v (m/s)")$
```

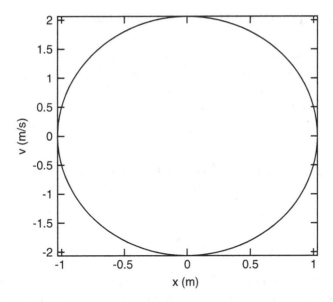

Fig. 4.5 Phase space trajectory for a simple harmonic oscillator

This plot shows that the simple harmonic oscillator follows an elliptical path in phase space, beginning at the point $(x = 1, v = 0.5)$ and moving clockwise. The ellipse is centered at the origin and has a length of $2A$ along the x-axis, and a length of $2A\omega_0$ along the v-axis. Changing the initial conditions may change the size of the ellipse, or it may change the point on the ellipse at which the motion begins, but the motion will always follow a clockwise, elliptical path in phase space.

We could have predicted that the phase space path for a simple harmonic oscillator would be an ellipse by considering conservation of energy. If the oscillator has total mechanical energy E, then

$$\frac{1}{2}kx^2 + \frac{1}{2}mv^2 = E, \tag{4.8}$$

which we can rewrite as

$$\frac{x^2}{a^2} + \frac{v^2}{b^2} = 1, \tag{4.9}$$

where $a^2 = 2E/k$ and $b^2 = 2E/m$. Equation 4.9 is the standard form of the equation for an ellipse in the x–v plane, centered at the origin, with length $2a$ along the x-axis and length $2b$ along the v-axis. Since $E = (1/2)kx_0^2 + (1/2)mv_0^2$ we find that $a = A$ and $b = A/\omega_0$, in agreement with our analysis above.

4.3   Two-Dimensional Harmonic Oscillator

Now that we have solved the one-dimensional harmonic oscillator, we extend our solution to the case of a two-dimensional harmonic oscillator. We assume a Hooke's Law type of force, but we do not necessarily assume that the force constants will be the same along each axis. So the force is given by

$$\vec{F} = -k_x x \hat{x} - k_y y \hat{y}, \tag{4.10}$$

where k_x and k_y are the force constants in the x- and y-directions, respectively. We can write Newton's Second Law as two separate equations:

$$m\ddot{x} = -k_x x,$$
$$m\ddot{y} = -k_y y. \tag{4.11}$$

Each of these equations is identical to the equation of motion for a one-dimensional harmonic oscillator, and therefore the solutions will be of the same form as for the one-dimensional case:

$$x(t) = x_0 \cos(\omega_x t) + (v_{x0}/\omega_x) \sin(\omega_x t), \text{ and}$$
$$y(t) = y_0 \cos(\omega_y t) + (v_{y0}/\omega_y) \sin(\omega_y t), \tag{4.12}$$

where $\omega_x = \sqrt{k_x/m}$ and $\omega_y = \sqrt{k_y/m}$.

Let's take a look at the motion of this two-dimensional harmonic oscillator. First we consider an *isotropic* harmonic oscillator for which $k_x = k_y = k$. The code below generates a plot of the trajectory of an isotropic oscillator initially displaced from the origin in both the x and y directions. The resulting plot, illustrating the motion of the oscillator in the x–y plane, is shown in Fig. 4.6.

```
(%i) x(t):=x0*cos(%omega[X]*t)+(vx0/%omega[X])*
     sin(%omega[X]*t)$
     y(t):=y0*cos(%omega[Y]*t)+(vy0/%omega[Y])*
     sin(%omega[Y]*t)$
     [xExpr1: subst([x0=1,y0=1,vx0=0,vy0=0,
     omega[X]=1,omega[Y]=1], x(t)),
     yExpr1: subst([x0=1,y0=1,vx0=0,vy0=0,
     %omega[X]=1,%omega[Y]=1], y(t))]$
     wxdraw2d(nticks=100,parametric(xExpr1,yExpr1,
        t,0,10),xlabel="x (m)", ylabel="y (m)",xaxis=true,
        yaxis=true)$
```

The motion of the oscillator is confined to a line containing the initial point and the origin. This is not surprising if we consider that the force on our isotropic oscillator is always directed toward the origin (since we can rewrite the force as $\vec{F} = -k\vec{r}$, where $\vec{r} = x\hat{x} + y\hat{y}$). When we release the oscillator from rest it will be

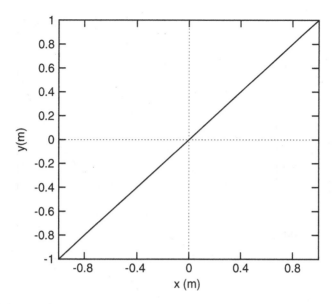

Fig. 4.6 Trajectory of a two-dimensional isotropic oscillator released from rest at $x_0 = 1$ m and $y_0 = 1$ m

pulled directly toward the origin. It will then overshoot the origin and continue to the point opposite its initial position. It will continue to oscillate between these two points indefinitely.

What happens if we give the oscillator an initial velocity? If we revise the code above to give the oscillator an initial x-component of velocity of 1 m/s, while keeping the other initial values the same ($x_0 = y_0 = 1$ m and $v_{y0} = 0$), the resulting plot is that shown in Fig. 4.7.[1]

Giving the oscillator an initial velocity in the x-direction changes the path from a line to an ellipse. The motion is still periodic: it repeats this same elliptical path over and over. It turns out that these are the only two types of paths that an isotropic 2D harmonic oscillator can follow: a line or an ellipse. Now let's look at an *anisotropic* case.

First consider a case where $\omega_y = 3\omega_x$. We release the oscillator from rest at a point displaced from the origin along both the x- and y-directions. Figure 4.8 shows the resulting plot with $\omega_x = 1$ rad/s, $\omega_y = 3$ rad/s, $x_0 = y_0 = 1$ m, and no initial velocity.

The path is no longer a straight line, as it was for the corresponding isotropic case, but it is still a one-dimensional curve. The oscillator moves along this curve and back again, repeating the same motion over and over. Note how the path illustrates the difference in ω_x and ω_y. While the oscillator completes a single oscillation along the x-direction, it also completes three full oscillations along the y-direction.

---

[1]The commands, essentially identical to those above, are omitted.

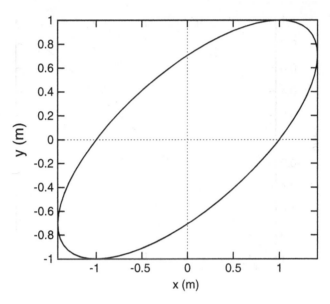

Fig. 4.7 Trajectory of an isotropic oscillator with $x_0 = y_0 = 1$ m, $v_{x0} = 1$ m/s, and $v_{y0} = 0$

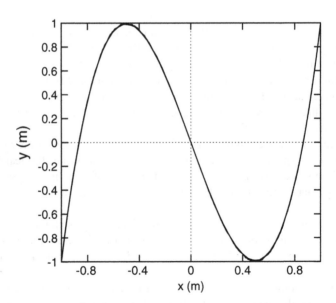

Fig. 4.8 Trajectory of an anisotropic oscillator ($\omega_x = 1$ rad/s, $\omega_y = 3$ rad/s) released from rest at $x_0 = y_0 = 1$ m

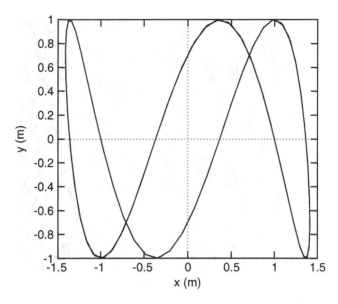

Fig. 4.9 Trajectory of an anisotropic oscillator ($\omega_x = 1$ rad/s, $\omega_y = 3$ rad/s) with initial conditions $x_0 = y_0 = 1$ m, $v_{x0} = 1$ m/s, and $v_{y0} = 0$

Figure 4.9 shows what happens if we add an initial velocity in the x-direction to our anisotropic oscillator. The resulting motion is no longer one-dimensional (as it was when the oscillator was released from rest), nor is it an ellipse (as for the corresponding isotropic case), but it is still periodic. In fact, we can still see that the particle completes three y-oscillations for every x-oscillation. The period of the motion in the y-direction is exactly one-third of the period of motion in the x-direction, so after each x-oscillation (and three y-oscillations) the system returns to its initial state and the motion repeats.

In fact, the motion will be periodic as long as $n\omega_x = m\omega_y$ for some integers n and m because every m oscillations in x will correspond to exactly n oscillations in y. We can rewrite our condition as $\omega_x/\omega_y = m/n$. Another way to state this condition is that the ratio of the two frequencies is a rational number. In this case we say the frequencies are *commensurable*.

If the frequencies are *incommensurable* (that is, their ratio is not a rational number) then the motion of the oscillator will not be periodic. Let's consider an example where $\omega_y = \pi\omega_x$, so that the ratio of the frequencies is the irrational number π. Figure 4.10 illustrates this case, showing the first 50 oscillations in the y-direction.

The particle now oscillates back and forth along the x- and y-directions, but without ever quite returning to its starting point. Thus, the motion never really repeats. This type of motion, consisting of two periodic motions with incommensurable periods, is called *quasiperiodic*. We get a better idea of the long-term behavior of this system if we plot the motion for a longer time. Figure 4.11 shows this system's behavior over 150 oscillations in the y-direction.

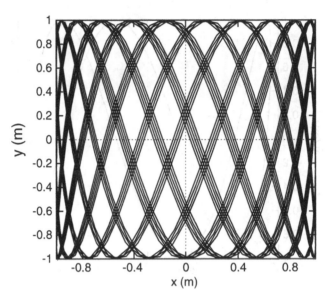

Fig. 4.10 Trajectory for an incommensurate anisotropic oscillator ($\omega_x = 1$ rad/s, $\omega_y = \pi$ rad/s) with $x_0 = y_0 = 1$ m and no initial velocity. The plot shows the motion during the first 50 oscillations in the y-direction (from $t = 0$ to $t = 100$ s)

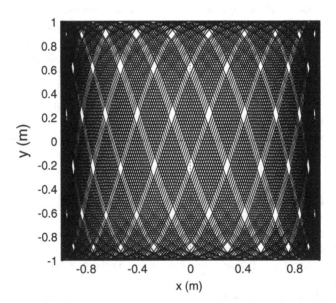

Fig. 4.11 The same as in Fig. 4.10, except over 150 oscillations in the y-direction (from $t = 0$ to $t = 300$ s)

This path is gradually filling a rectangular region of the x–y plane. Eventually the path will pass arbitrarily close to any point in this region. Such motion is said to be *ergodic*.

4.4 Damped Harmonic Oscillator

So far the oscillators we have considered have all been idealized systems that will oscillate forever at a constant amplitude, never losing energy. Realistic mechanical oscillators do lose energy due to damping forces, such as internal friction within a spring. This section investigates the motion of a damped harmonic oscillator, in which the Hooke's Law force on our oscillator is joined by a linear damping force of the form $\vec{f} = -b\vec{v}$, where $\vec{v}$ is the velocity of the oscillator.

We can write Newton's Second Law for this system as

$$m\ddot{x} = -kx - b\dot{x}. \tag{4.13}$$

Dividing by the mass, and defining the constant $\beta = b/(2m)$, we have

$$\ddot{x} = -\omega_0^2 x - 2\beta\dot{x}, \tag{4.14}$$

where $\omega_0 = \sqrt{k/m}$ is the natural frequency of the undamped oscillator.

We use *Maxima* to solve the differential equation in Eq. 4.14, but as we see the solution depends on the relation between β and ω_0. For now, let us assume that $\beta < \omega_0$, a situation we will refer to as an "underdamped" oscillator.

```
(%i) assume(%beta>0,
        %omega[0]>0, %beta<%omega[0])$
    atvalue(x(t),t=0,x0)$
    atvalue('diff(x(t),t),t=0,v0)$
    eq1:'diff(x(t),t,2)=-%omega[0]^2*x(t)-2*%beta*
        'diff(x(t),t)$        sol:desolve(eq1,x(t));
```

$$(\%o)\ \ x(t) = e^{-\beta t}\left(\frac{\sin\left(\sqrt{\omega_0^2-\beta^2}\,t\right)(2(2\beta x0+v0)-2\beta x0)}{2\sqrt{\omega_0^2-\beta^2}} + \cos\left(\sqrt{\omega_0^2-\beta^2}\,t\right)x0\right)$$

We can simplify this solution by defining a new constant ω_1, such that $\omega_1^2 = \omega_0^2 - \beta^2$. Note that our solution above consists of the sum of two trigonometric functions that oscillate with frequency ω_1, but with that entire sum multiplied by an exponential factor that decreases with time.

```
(%i) x(t):=exp(-%beta*t)*(((v0+%beta*x0)/%omega[1])*
        sin(%omega[1]*t)+x0*cos(%omega[1]*t));
    x0:1$ v0:0$ %omega[0]:1$ %beta:0.1$
    %omega[1]:sqrt(%omega[0]^2-%beta^2)$
```

$$(\%o)\ \ x(t) := \exp\left((-\beta)\,t\right)\left(\frac{v0+\beta\,x0}{\omega_1}\sin(\omega_1 t) + x0\cos(\omega_1 t)\right)$$

Fig. 4.12 Plots of position
(*top left*) and velocity (*top
right*) as a function of time, as
well as the phase space
trajectory (*bottom left*) and
energy per mass as a function
of time (*bottom right*) for an
underdamped harmonic
oscillator

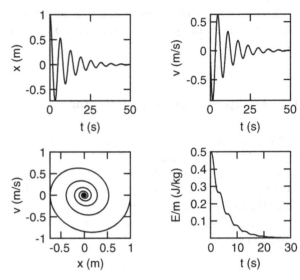

4.4.1   Underdamped Oscillators

To see what the motion looks like we plot the position and velocity as a function of
time, as well as the motion in phase space (v versus x). We assume that the oscillator
starts from rest, but displaced from the origin, with $\omega_0 = 1$ rad/s and $\beta = 0.1$ in the
same units. The code below shows how to display all three of these plots, as well
as a plot of energy versus time (discussed below). Figure 4.12 shows the resulting
plots. The top two panels show the values of x and v as a function of time. The
bottom left panel shows the phase space trajectory for this oscillator. The final panel
will be discussed below.

```
(%i) xpathUD:gr2d(explicit(x(t),t,0,50),
        xlabel="t (s)", ylabel="x (m)",
        xtics=25,ytics=.5)$
     v(t):="(diff(x(t),t))$
     vpath UD: gr2d(explicit(v(t),t,0,50),
        xlabel="t (s)", ylabel="v (m/s)",
        ytics=.5,xtics=25)$
     phaseUD: gr2d(nticks=400,
        parametric(x(t),v(t),t,0,50), xlabel="x (m)",
        ylabel="v (m/s)",yrange=[-1,1],xtics=.5)$
     EmpathUD: gr2d(explicit(0.5*%omega[0]^2*x(t)^2+
        0.5*v(t)^2,t,0,30), xlabel="t (s)",
        ylabel="E/m (J/kg)",ytics=.1,xtics=10)$
     wxdraw(xpathUD, vpathUD,phaseUD,EmpathUD,
        dimensions=[640,480], columns=2)$
```

The top two panels of Fig. 4.12 show that both the position and velocity graphs
are oscillating functions, but with amplitudes that decay with time. The path in phase

space is a spiral that gradually approaches the origin. All of these plots make it clear that after a long time this oscillator will come to rest at $x = 0$. The loss of motion occurs because the damping force is removing energy from the system. Once all of the mechanical energy has been removed, the system will come to rest at the equilibrium point.

The mechanical energy in this system is given by

$$E = \frac{1}{2}mv^2 + \frac{1}{2}kx^2. \tag{4.15}$$

We can examine what happens to the mechanical energy in this system by plotting the total mechanical energy per unit mass,

$$E/m = \frac{1}{2}v^2 + \frac{1}{2}\omega_0^2 x^2. \tag{4.16}$$

as a function of time. The resulting plot is shown in the bottom right of Fig. 4.12.

The energy decreases rapidly at first, because the oscillator is moving rapidly and therefore the damping force ($\vec{f} = -b\vec{v}$) is strong. As more energy is removed from the system the oscillator's motion slows, and the damping force is not as strong. Thus, the energy decrease is generally less rapid at later times. The oscillations in the energy versus time curve occur because the damping force has little effect when the oscillator is near its endpoints (and moving slowly) but has its greatest effect when the oscillator is passing through the equilibrium point (and moving rapidly).

4.4.2 Overdamped Oscillators

Now that we understand the motion of the underdamped harmonic oscillator, we can consider the "overdamped" case when $\beta > \omega_0$. We can use *Maxima* again to solve Eq. 4.14 for this new case.

```
(%i) kill(all)$
     assume(%beta>0,%omega[0]>0, %beta>%omega[0])$
     atvalue(x(t),t=0,x0)$ atvalue('diff(x(t),t),t=0,v0)$
     eq1:'diff(x(t),t,2)=-%omega[0]^2*x(t)-2*%beta*
         'diff(x(t),t)$          sol:desolve(eq1,x(t));
```

$$(\%o)\ x(t) = e^{-\beta t}\left(\frac{\sinh\left(\sqrt{\beta^2-\omega_0^2}\,t\right)(2\,(2\,\beta\,x0+v0)-2\,\beta\,x0)}{2\,\sqrt{\beta^2-\omega_0^2}} + \cosh\left(\sqrt{\beta^2-\omega_0^2}\,t\right)x0\right)$$

This solution looks like the solution for the underdamped case, except that the trigonometric functions have been replaced by hyperbolic functions (sinh rather than sin, and so on). We can simplify the result by defining the new constant ω_2, such that $\omega_2^2 = \beta^2 - \omega_0^2$. Before we construct plots of our solution, it helps to rewrite the solution in terms of exponential functions (rather than hyperbolic functions) and expand the multiplication of the various exponential factors. To achieve this, we use *Maxima*'s exponentialize command.

```
(%i)  x(t):=exp(-%beta*t)*(sinh(%omega[2]*t)*
         (%beta*x0+v0)/%omega[2]+ x0*cosh(%omega[2]*t))$
      expand(exponentialize(x(t)));
```

$$(\%o)\quad \frac{\beta e^{\omega_2 t-\beta t}x0}{2\omega_2} + \frac{e^{\omega_2 t-\beta t}x0}{2} - \frac{\beta e^{-\beta t-\omega_2 t}x0}{2\omega_2} + \frac{e^{-\beta t-\omega_2 t}x0}{2} + \frac{e^{\omega_2 t-\beta t}v0}{2\omega_2} - \frac{e^{-\beta t-\omega_2 t}v0}{2\omega_2}$$

The result has six different terms, and each term contains an exponential factor. Three of the terms have an $e^{-(\beta-\omega_2)t}$ factor, and three have an $e^{-(\beta+\omega_2)t}$ factor. Since $(\beta + \omega_2) > (\beta - \omega_2)$ we can see that the terms with the $e^{-(\beta+\omega_2)t}$ factor will decay more rapidly than the other terms. The long-term behavior of this oscillator will be governed by the slowly decaying terms, which decay exponentially with a rate $(\beta - \omega_2)$ (unless the initial conditions are such that the slowly decaying terms cancel each other out).

To visualize the motion of this overdamped oscillator, we construct plots of position and velocity as a function of time, as well as the phase space trajectory and the energy (per unit mass) as a function of time. Figure 4.13 shows the results.[2] We assume the same initial conditions that were used above for the underdamped case, but this time we will use $\beta = 5\ \text{s}^{-1}$.

This "oscillator" moves steadily toward the origin, asymptotically approaching the origin as $t \to \infty$. The velocity of the oscillator initially becomes negative, but then this velocity too decays away, approaching zero in the limit $t \to \infty$.

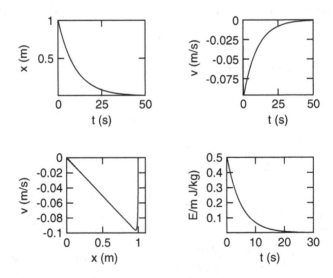

Fig. 4.13 Plots of position (*top left*) and velocity (*top right*) as a function of time, as well as the phase space trajectory (*bottom left*) and energy per mass as a function of time (*bottom right*) for an overdamped harmonic oscillator

---

[2]The commands, essentially the same as those used to generate Fig. 4.12, are omitted.

The phase space path is particularly interesting because it illustrates that there are two distinct phases to the motion of this oscillator. In the first phase the oscillator moves toward the equilibrium point and picks up a negative velocity, much as we might expect for an undamped harmonic oscillator. However, the oscillator then makes a sharp turn in phase space and subsequently approaches the origin along a straight line path.

We can better understand this two part motion by referring to the exponential factors in the solution for $x(t)$ discussed above. For the parameter values we have chosen, the values of β and ω_2 are very similar ($\beta = 5$ s^{-1}, while $\omega_2 = \sqrt{24} \approx 4.9$ s^{-1}). Therefore, the terms that decay at the rate $(\beta + \omega_2)$ do so very rapidly, while the terms that decay at the rate $(\beta - \omega_2)$ do so much more slowly. The second phase of the motion, in which the oscillator gradually approaches the origin along a straight line in phase space, corresponds to the slowly decaying terms, after the rapidly decaying terms have already vanished.

Figure 4.13 shows that, as with the underdamped oscillator, the energy rapidly decreases at the beginning, but decreases more gradually at later times. In this case there are no oscillations in the energy curve, because this oscillator doesn't actually oscillate! It just gradually approaches the origin from one side. Note that it takes about as long for the energy to vanish from this overdamped oscillator as it did from the underdamped oscillator we examined above. Although the damping force is much stronger in the overdamped case, the damping force actually prevents the oscillator from moving quickly back to its equilibrium point. Thus, even though the damping force quite effectively removes kinetic energy from the system, it actually slows the removal of potential energy.

4.4.3 Critical Damping

Finally, we examine the case of "critical damping" when $\beta = \omega_0$. Again, we use *Maxima* to solve Eq. 4.14, but this time replacing ω_0 with β.

```
(%i)  kill(all)$ atvalue(x(t),t=0,x0)$
      atvalue('diff(x(t),t),t=0,v0)$ eq1:'diff(x(t),t,2)=
        -%beta^2*x(t)-2*%beta*'diff(x(t),t)$
      sol:ratsimp(desolve(eq1,x(t)));
(%o)  x(t) = e^{-βt} ((β t + 1) x0 + t v0)
```

The solution consists of a factor that is linear in time $((tv_0 + (\beta t + 1)x_0)$ and an exponential decay factor $(e^{-\beta t})$. To examine the behavior of this critically damped oscillator we will again construct plots of $x(t)$, $v(t)$, v versus x, and E/m as a function of t. We will use the same initial conditions as we used for the previous oscillators, but this time with $\beta = 1$ s^{-1}. The resulting plots are shown in Fig. 4.14.

In general we see that the behavior is similar to that of the overdamped oscillator. The oscillator moves directly toward the origin where it comes to rest. The velocity initially becomes negative and then decays to zero. The path through phase space

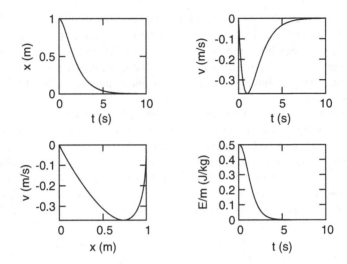

Fig. 4.14 Plots of position (*top left*) and velocity (*top right*) as a function of time, as well as the phase space trajectory (*bottom left*) and energy per mass as a function of time (*bottom right*) for a critically damped harmonic oscillator

begins like that of an undamped oscillator, but then curves and moves toward the origin where it ends (although the curve is much more gradual for the critically damped oscillator than for the overdamped oscillator). However, note the shorter time scale in the graphs for the critically damped oscillator. The different time scales become apparent when we consider a plot of energy versus time for the case of critical damping, as shown in the bottom right of Fig. 4.14.

Note how rapidly the energy decays to zero. Critical damping leads to the fastest possible removal of energy, and thus the motion of an oscillator will die out quickest when the damping is critical. This explains why shock absorbers on vehicles (cars, bicycles) aim for critical damping. When you hit a bump, you want your wheel to return to its proper position as quickly as possible and stay there. An underdamped shock will cause the wheel to oscillate before returning to its equilibrium location, while an overdamped shock will return to equilibrium too slowly (so that the shock may not be ready for the next impact).

4.5 Driven Damped Harmonic Oscillator

In the previous section we saw that a harmonic oscillator subject to a damping force will eventually cease its motion. But what would happen if there was an external force acting on the oscillator to keep it in motion? In this section we examine the motion of a harmonic oscillator that is subject to damping and also to an external force that varies sinusoidally in time.

We consider external forces that vary periodically in time because such forces will keep the oscillator oscillating—unlike, for example, a constant force which would only result in the oscillator coming to rest at a displaced equilibrium point. We consider sinusoidally varying forces first, because such forces are easiest to deal with, mathematically. In the next section we will look at how to handle non-sinusoidal periodic forces.

The external force that drives the oscillator takes the form $F\cos(\omega t)$, where ω is the angular frequency of the driving force and F is the amplitude (maximum magnitude) of the driving force. Newton's Second Law for this system is

$$m\ddot{x} = -kx - b\dot{x} + F\cos(\omega t). \tag{4.17}$$

Dividing through by the mass and defining the new constant $f = F/m$ we have

$$\ddot{x} = -\omega_0^2 x - 2\beta\dot{x} + f\cos(\omega t), \tag{4.18}$$

where $\omega_0 = \sqrt{k/m}$ and $\beta = b/(2m)$ as for the damped harmonic oscillator. We can use *Maxima* to solve Eq. 4.18, assuming an underdamped oscillator ($\beta < \omega_0$).[3]

```
(%i) assume(%beta>0, %beta<%omega[0], %omega>0)$
     eq1:'diff(x(t),t,2)=-%omega[0]^2*x(t)-2*%beta*
         'diff(x(t),t)+f*cos(%omega*t)$
     atvalue(x(t),t=0,x0)$ atvalue('diff(x(t),t),t=0,v0)$
     sol:desolve(eq1,x(t));
```

$$(\%o)\quad x(t) = e^{-\beta t}\Bigg(\Big(\sin\big(\sqrt{\omega^2-\beta^2}\,t\big)\Big)\,\Bigg($$

$$\frac{2\left(\left(2\beta\omega^4+\left(8\beta^3-4\omega_0^2\beta\right)\omega^2+2\omega_0^4\beta\right)x0+\left(\omega^4+\left(4\beta^2-2\omega_0^2\right)\omega^2+\omega_0^4\right)v0-2\omega_0^2\beta f\right)}{\omega^4+\left(4\beta^2-2\omega_0^2\right)\omega^2+\omega_0^4} -$$

$$\frac{2\beta\left(\left(\omega^4+\left(4\beta^2-2\omega_0^2\right)\omega^2+\omega_0^4\right)x0+\left(\omega^2-\omega_0^2\right)f\right)}{\omega^4+\left(4\beta^2-2\omega_0^2\right)\omega^2+\omega_0^4}\Bigg)\Bigg)/\left(2\sqrt{\omega_0^2-\beta^2}\right)+$$

$$\frac{\cos\left(\sqrt{\omega_0^2-\beta^2}\,t\right)\left(\left(\omega^4+\left(4\beta^2-2\omega_0^2\right)\omega^2+\omega_0^4\right)x0+\left(\omega^2-\omega_0^2\right)f\right)}{\omega^4+\left(4\beta^2-2\omega_0^2\right)\omega^2+\omega_0^4}\Bigg)+$$

$$\frac{2\beta\omega f\sin(\omega t)}{\omega^4+\left(4\beta^2-2\omega_0^2\right)\omega^2+\omega_0^4} - \frac{\left(\omega^2-\omega_0^2\right)f\cos(\omega t)}{\omega^4+\left(4\beta^2-2\omega_0^2\right)\omega^2+\omega_0^4}$$

The solution is quite complicated, but note that all but the final two terms (the two on the bottom output row) contain a factor of $e^{-\beta t}$. Therefore we can expect the first part of the solution to decay away, and at long times the solution will be dominated by the final two terms, both of which are sinusoidal functions of time with angular frequency ω (the frequency of the driving force). Before we focus on this "steady state" portion of the solution, though, let's examine the full solution by loading the above result into a new function (note the use of double quotes and parentheses in the code below) and constructing a plot of position versus time. To illustrate this process, we assume $\omega_0 = 1$ rad/s, $\omega = 1.3$ rad/s, $\beta = 0.1$ s^{-1}, $f = 1$ N/kg, $x_0 = 1$ m, and $v_0 = 1$ m/s. Figure 4.15 shows the resulting plot.

---

[3] The steady state portion of the solution, which is the part we are most interested in, is the same whether the oscillator is underdamped, overdamped, or critically damped.

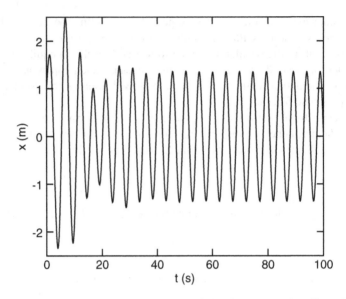

Fig. 4.15 Position as a function of time for a sinusoidally driven, damped oscillator. Parameters and initial conditions are given in the text

```
(%i) x(t):="(rhs(sol))$ %omega[0]:1$ %beta:0.1$
      %omega:1.3$ f:1$ x0:1$ v0:1$
   wxdraw2d(explicit(x(t),t,0,100),yrange=[-2.5,2.5],
      xlabel="t (s)",ylabel="x (m)")$
```

We see that after some initial transient behavior the system settles into a steady sinusoidal oscillation with a constant amplitude and well-defined period. It is this "steady state" solution that is represented by the final two terms in our solution above. We can write this steady state solution as

$$x_{ss}(t) = \frac{f(2\beta\omega \sin(\omega t) + (\omega_0^2 - \omega^2)\cos(\omega t))}{(\omega_0^2 - \omega^2)^2 + (2\beta\omega)^2}. \tag{4.19}$$

This steady state solution does not depend on the initial position or velocity. We can construct a plot of our steady state solution using the code below. The resulting plot is shown in Fig. 4.16.

```
(%i) xss(t):=f*(2*%beta*%omega*sin(%omega*t)+
      (%omega[0]^2- %omega^2)*cos(%omega*t))/
      ((%omega[0]^2-%omega^2)^2+(2*%beta*%omega)^2)$
   %omega[0]:1$ %beta:0.1$ %omega:1.3$ f:1$
   wxdraw2d(explicit(xss(t),t,0,100),yrange=
      [-1.5,1.5], xlabel="t (s)",
      ylabel="x_{ss} (m)")$
```

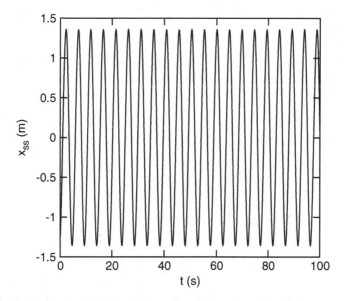

Fig. 4.16 Steady state solution for the driven, damped harmonic oscillator shown in Fig. 4.15

Comparing the plot of the steady state solution and the full solution shows that after about $t = 40$ s the two solutions are indistinguishable. It is also clear from this plot that the steady state solution is a simple sinusoidal function. In fact, we can write the steady state solution as

$$x_{ss}(t) = A\cos(\omega t - \delta). \tag{4.20}$$

To see how to rewrite the solution in this way, we can expand the new form of the solution given in Eq. 4.20.

```
(%i) kill(all)$   trigexpand(A*cos(%omega*t-%delta));
(%o) [(sin(δ) sin(ω t) + cos(δ) cos(ω t)) A
```

To match with the form given in Eq. 4.19 we must have

$$A\sin\delta = \frac{2f\beta\omega}{(\omega_0^2 - \omega^2)^2 + (2\beta\omega)^2}, \tag{4.21}$$

and

$$A\cos\delta = \frac{f(\omega_0^2 - \omega^2)}{(\omega_0^2 - \omega^2)^2 + (2\beta\omega)^2}. \tag{4.22}$$

Fig. 4.17 A right triangle
illustrating the relation
between parameters in
Eqs. 4.19 and 4.20

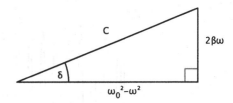

Adding the squares of Eqs. 4.21 and 4.22 we find that

$$A^2 = \frac{f^2}{(\omega_0^2 - \omega^2)^2 + (2\beta\omega)^2},$$ (4.23)

and dividing Eq. 4.21 by Eq. 4.22 we find that

$$\delta = \tan^{-1}\left(\frac{2\beta\omega}{\omega_0^2 - \omega^2}\right).$$ (4.24)

The triangle in Fig. 4.17 illustrates the relation between the phase angle δ and the parameters ω, β, and ω_0. The amplitude of the steady state solution is then $A = f/C$, where C is the hypotenuse of the triangle in Fig. 4.17. If we define new dimensionless constants $q = \omega/\omega_0$ and $p = \beta/\omega_0$, then we have

$$A^2 = \frac{f^2}{\omega_0^4} \frac{1}{(2qp)^2 + (1 - q^2)^2}.$$ (4.25)

The first factor on the right-hand side of Eq. 4.25 does not depend on q or p, and thus does not depend on ω or β. We will refer to the other factor as the relative square amplitude and we can plot this relative square amplitude function versus q to see how the amplitude of the steady state solution depends on the driving frequency $\omega = q\omega_0$. The code below constructs plots of the relative square amplitude as a function of q for two different values of p: $p = 0.1$ and $p = 0.3$. The resulting plots are shown in Fig. 4.18.

```
(%i) SqAmp(q,p):=1/((2*q*p)^2+(1-q^2)^2)$
     withp01: gr2d(title="p=0.1",explicit(
     SqAmp(q,0.1),q,0,3),xlabel="q", ylabel=
     "relative square amplitude",xtics=1,ytics=10)$
     withp03: gr2d(title="p=0.3",explicit(
     SqAmp(q,0.3),q,0,3),xlabel="q",
     xtics=1,ytics=1)$
     wxdraw(withp01,withp03,columns=2)$
```

The first panel of Fig. 4.18 shows that when $p = 0.1$ the relative square amplitude of the steady state solution peaks near $q = 1$ or, equivalently, near $\omega = \omega_0$. This phenomenon is known as *resonance*. The oscillator responds much more strongly to a driving force with a frequency near that of the oscillator's natural frequency.

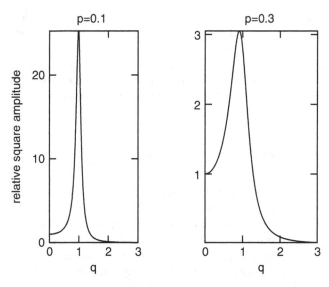

Fig. 4.18 Relative square amplitude of the driven, damped harmonic oscillator as a function of $q = \omega/\omega_0$, for two values of $p = \beta/\omega_0$

However, close inspection of the plot reveals that the peak occurs at a value of q slightly less than one. What happens if we change the value of p? The second panel of Fig. 4.18 shows the resonance peak for $p = 0.3$.

Note that the peak is much wider for this larger value of p, which corresponds to a larger value of β. Damping forces widen the resonance peak for the oscillator. Also, the location of the peak has shifted toward a lower q value. We can determine the exact location of the peak by using calculus to find the value of q that maximizes the relative square amplitude function.

```
(%i) solve("diff(SqAmp(q,p),q)=0,q);
(%o) [q = -√(1-2p²), q = √(1-2p²), q = 0]
```

The peak in the relative square amplitude occurs when $q = \sqrt{1 - 2p^2}$, or when $\omega = \sqrt{\omega_0^2 - 2\beta^2}$ (the other two solutions are not physically relevant). We can evaluate the relative square amplitude at this resonance frequency.

```
(%i) SqAmp(sqrt(1-p^2),p);        (%o) 1/(p⁴+4p²(1-p²))
```

Multiplying by the factor of f^2/ω_0^4 from Eq. 4.25 we find that the square amplitude at resonance is

$$A_{res}^2 = \frac{f^2}{4\beta^2\omega_0^2 - 3\beta^4}. \tag{4.26}$$

For weak damping ($\beta \ll \omega_0$) this result reduces to $A_{res} \approx f/(2\beta\omega_0)$. Stronger damping forces not only shift the resonance frequency farther below the natural frequency of the oscillator, but they also reduce the amplitude of the steady state solution at resonance. Therefore, resonance behavior will be strongest, and narrowly confined to frequencies near ω_0, when there is minimal damping. When there is strong damping the resonance effect will be much less noticeable.

4.6 Non-sinusoidal Driving Forces

The previous section considered a harmonic oscillator subject to a sinusoidal driving force. This section shows how to solve for the motion of a harmonic oscillator subject to a non-sinusoidal (but still periodic) driving force. We focus on the special case of an oscillator driven by a periodic series of square pulses, but the methods used to solve this problem apply to any periodic driving force.

We begin by defining our non-sinusoidal driving force. We consider only periodic forces such that $F(t + T) = F(t)$, where T is the period of the function. We want this driving force to consist of a series of square-shaped pulses, so that the external driving force on the oscillator is either zero or a constant nonzero value. The force will be "on" for a time dt and then "off" for the remainder of the period T. The angular frequency of the pulses is $\omega = 2\pi/T$.

We define a function $s(t)$ that produces such a series of square pulses using *Maxima*'s floor function. The floor function, sometimes known as the greatest integer function, returns the greatest integer that is less than its argument. The code below shows how to use the difference of two floor functions to construct and plot a series of square pulses with period $T = 2\pi$ s (so $\omega = 1$ rad/s) and pulse width $dt = 1$ s. Figure 4.19 shows the resulting plot.

```
(%i) s(t):=floor(%omega*(t+dt/2)/(2*%pi))-
       floor(%omega*(t-dt/2)/(2*%pi));
     %omega:1$ dt:1$
     wxdraw2d(explicit(s(t),t,-10,10),
       yrange=[-1,2],xlabel="t (s)",ylabel="s")$
```

We can write Newton's Second Law for our driven oscillator as

$$\ddot{x} = -\omega_0^2 x - 2\beta\dot{x} + f(t), \qquad (4.27)$$

where $f(t) = F(t)/m$. We can solve this differential equation numerically using *Maxima*'s rk command, with $f(t) = f_{max}s(t)$ (where f_{max} is just the force value for the square pulses divided by the mass of the oscillator). The code below show how to generate and plot the solution for $\omega_0 = 1$ rad/s, $\beta = 0.5$ s^{-1}, and $f_{max} = 1$ N/kg. The resulting plot is shown in Fig. 4.20.

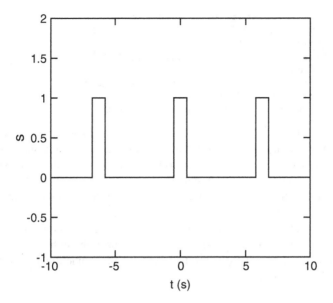

Fig. 4.19 A periodic series of square wave pulses

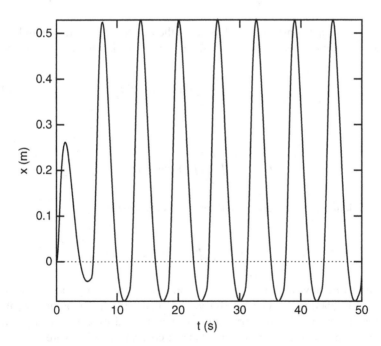

Fig. 4.20 Numerical solution of $x(t)$ for the harmonic oscillator driven by a periodic series of square pulses

```
(%i) %omega[0]:1$ %beta:0.5$ f:1$ data1:
     rk([v,-%omega[0]^2*x-2*%beta*v+f*s(t)],
        [x,v],[0,0],[t,0,50,0.1])$
     xvt:makelist([data1[i][1],data1[i][2]],i,1,
        length(data1))$
     wxdraw2d(xlabel="t (s)",ylabel="x (m)",
        xaxis=true, point_size=0,
        points_joined=true,points(xvt))$
```

Figure 4.20 shows that the oscillator settles into a periodic pattern of steady state motion after some initial transient behavior. However, the steady state oscillations are not sinusoidal. The natural oscillations of the harmonic oscillator are sinusoidal, but the driving force is non-sinusoidal. The combination of the two forces produce a periodic motion that is does not match the shape of the natural oscillations or the driving force.

Our numerical solution is helpful, but an analytical solution might be even better. We can construct an analytical solution to our problem by taking advantage of Fourier's theorem, which states that any periodic function $f(t)$ with period $T = 2\pi/\omega$ can be written as

$$f(t) = \sum_{n=0}^{\infty} [a_n \cos(n\omega t) + b_n \sin(n\omega t)], \qquad (4.28)$$

where a_n and b_n are constants that depend on the function $f(t)$. The sum in Eq. 4.28 is known as a *Fourier series* and the coefficients in the series, for $n \geq 1$, are given by

$$a_n = \frac{2}{T} \int_{-T/2}^{T/2} f(t) \cos(n\omega t) dt, \qquad (4.29)$$

and

$$b_n = \frac{2}{T} \int_{-T/2}^{T/2} f(t) \sin(n\omega t) dt. \qquad (4.30)$$

The coefficients for $n = 0$ are treated separately: $b_0 = 0$ and

$$a_0 = \frac{1}{T} \int_{-T/2}^{T/2} f(t) dt. \qquad (4.31)$$

We illustrate how the Fourier series works by constructing the series for the square pulse function $s(t)$ defined above. Note that $s(t)$ is periodic on the interval $[-T/2, T/2]$, but it is zero everywhere on that interval except on $[-dt/2, dt/2]$, where it has the value 1. Therefore, the Fourier coefficients for $s(t)$ are

$$a_n = \frac{2}{T} \int_{-dt/2}^{dt/2} \cos(n\omega t) dt,$$

$$b_n = \frac{2}{T} \int_{-dt/2}^{dt/2} \sin(n\omega t) dt,$$

$$a_0 = \frac{1}{T} \int_{-dt/2}^{dt/2} dt, \tag{4.32}$$

and $b_0 = 0$ (where the expressions for a_n and b_n apply for $n \geq 1$). We can calculate these integrals using *Maxima*, replacing T with $2\pi/\omega$.

```
(%i) kill(all)$
     a0:%omega*integrate(1,t,-dt/2,dt/2)/(%pi)$
     a(n):=%omega*integrate(cos(n*%omega*t),t,
        -dt/2,dt/2)/%pi$
     b(n):=%omega*integrate(sin(n*%omega*t),t,
        -dt/2,dt/2)/%pi$ [a0, a(n), b(n)];
```
$(\%o)\ [\frac{\omega\, dt}{\pi}, \frac{2\sin(\frac{\omega\, dtn}{2})}{\pi\, n}, 0]$

We see that $b_n = 0$ for all n. This result follows from the fact that $s(t)$ is an even function, therefore its Fourier series representation must consist only of even functions (cosines) with no odd functions (sines) contributing. We can also see that a_n is inversely proportional to n, so the coefficients will be negligibly small for large values of n. This fact is illustrated in Fig. 4.21, which shows a plot of a_n versus n for the function $s(t)$.

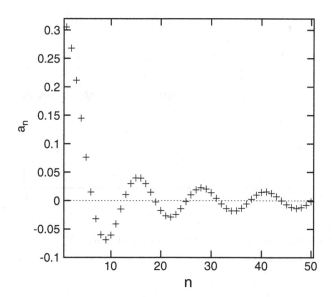

Fig. 4.21 Fourier coefficients a_n for the function $s(t)$ shown in Fig. 4.19

```
(%i) %omega:1$ dt:1$ fc:makelist(a(n),n,1,50)$
     wxdraw2d(points(fc),xrange=[0.5,50.5],
       yrange=[-0.1,0.32],xaxis=true),
       xlabel="n", ylabel="a_n"$
```

Recall that $a_0 = dt\omega/\pi \approx 0.318$. So a_0 and a_1 are of comparable size, but for higher values of n the value of a_n becomes generally smaller. Although the values of a_n oscillate about zero, the amplitude of these oscillations decreases steadily as n increases. This is a fortunate situation, since we would not be able to make much use of the infinite Fourier series if we really needed to calculate all of its terms. Thankfully, in this case and in most cases of practical interest, we can get accurate results by truncating the series at $n = n_{max}$. To illustrate how this works, we can construct the Fourier series for $s(t)$ up to $n = 10$ and plot the resulting function using the code below. The results are shown in Fig. 4.22.

```
(%i) ff(t):=a0+sum(a(n)*cos(n*%omega*t),n,1,nmax)$
     ff0(t):=''(subst([%omega=1,dt=1,nmax=10],ff(t)))$
     wxdraw2d(explicit(ff0(t),t,-10,10),yrange=[-1,2],
       xlabel="t (s)",ylabel="f" )$
```

Our truncated Fourier series with only 11 terms manages to reproduce the general shape of our square pulses, but the pulse tops are not quite flat, nor is the region in

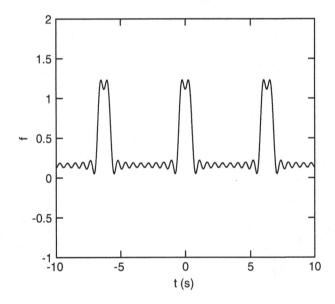

Fig. 4.22 Plot of the Fourier series for $s(t)$ using the first 11 terms

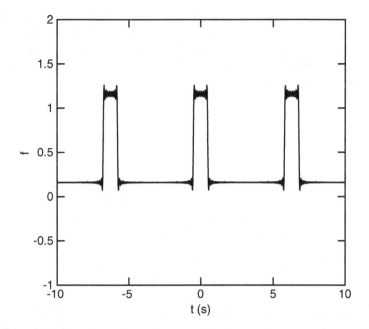

Fig. 4.23 Plot of the Fourier series for $s(t)$ using the first 51 terms

between the pulses. We now see what kind of improvement we get if we include terms up to $n = 50$. The plot using the first 51 terms is shown in Fig. 4.23.[4]

The Fourier series with 51 terms more accurately reproduces the square pulses, as Fig. 4.23 shows. The tops of the pulses are relatively flat, and the region between the pulses is quite flat. However, we see that the Fourier series tends to overshoot at the points of discontinuity in the $s(t)$ function, resulting in spikes that rise above one and fall below zero at the edges of each pulse. This is a common effect in Fourier series for periodic functions with jump discontinuities. The effect is usually known as the Gibbs phenomenon, and it is responsible for "ringing artifacts" in the processing of signals (including digital images and audio).

In Sect. 4.5 we saw that a damped harmonic oscillator, driven by a periodic force $F\cos(\omega t)$, has a steady state solution of the form

$$x_{ss}(t) = A\cos(\omega t - \delta). \tag{4.33}$$

Likewise (although we will not derive the solution here) a driving force $F\sin(\omega t)$ gives rise to a steady state solution of the form

$$x_{ss}(t) = A\sin(\omega t - \delta). \tag{4.34}$$

---

[4]The commands, essentially the same as those used to produce 4.22, are omitted.

We have now seen that any periodic force can be written as a sum of sine and cosine forces with frequencies $n\omega$. Because Eq. 4.27 is linear in x, the principle of superposition tells us that the solution when $f(t)$ is the sum of several sinusoidal forces is just the sum of the solutions for each individual sinusoidal force. Therefore, the solution for any periodic $f(t)$ is

$$x_{ss}(t) = \sum_{n=0}^{\infty} A_n \cos(n\omega t - \delta_n) + B_n \sin(n\omega t - \delta_n), \qquad (4.35)$$

where

$$A_n = a_n / \sqrt{(\omega_0^2 - n^2\omega^2)^2 + (2\beta n\omega)^2},$$

$$B_n = b_n / \sqrt{(\omega_0^2 - n^2\omega^2)^2 + (2\beta n\omega)^2}, \qquad (4.36)$$

$$\delta_n = \tan^{-1}\left(\frac{2\beta n\omega}{\omega_0^2 - n^2\omega^2}\right),$$

and a_n and b_n are the Fourier coefficients from Eqs. 4.29, 4.30, and 4.31.

We can apply this solution to the case of the oscillator driven by the square pulse function $s(t)$. We have already computed the Fourier coefficients a_n for this function (recall that $b_n = 0$ for all n because $s(t)$ is an even function). We can now use Eq. 4.36 to compute A_n and δ_n and then construct our solution from Eq. 4.35, but using only a finite number of terms in the sum (for this example we will use the first 11 terms).

We can then plot this approximate solution and compare it to the numerical solution generated by the rk command above. (Note the use of atan2 in the code below. Using the atan2 function ensures that the phase angles δ_n come out in the correct quadrant. Using the regular atan function can result in incorrect phase angles. In this particular case, using atan rather than atan2 would generate a solution that is the negative of the correct solution.) The code to generate the solution and plot is shown below, and the resulting plot is shown in Fig. 4.24.

```
(%i) amp(n):=f*a(n)/sqrt((%omega[0]^2-n^2*%omega^2)^2
     + 4*%beta^2*n^2*%omega^2)$
     delta(n):=atan2(2*%beta*n*%omega,%omega[0]^2-
     n^2*%omega^2)$
     xf(t):=%omega*dt*f/(2*%pi*%omega[0]^2)+
     sum(amp(n)*cos(n*%omega*t-delta(n)),n,1,nmax)$
     %omega:1$ dt:1$
     %omega[0]:1$ f:1$ %beta:0.5$ nmax:10$
     wxdraw2d(explicit(xf(t),t,0,50),xlabel="t (s)",
     ylabel="x (m)")$
```

Comparing Fig. 4.24 to Fig. 4.20 shows that the two plots disagree initially, but agree very well for later times. Since Eq. 4.35 provides only the steady state solution, ignoring the transient motion at the beginning, we don't expect it to match the actual

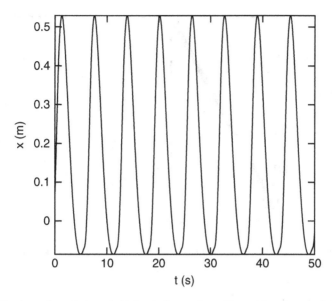

Fig. 4.24 Fourier series solution of $x(t)$ for the harmonic oscillator driven by a periodic series of square pulses. The first 11 terms of the Fourier series were used to construct this solution

motion of the system for small times. But at large times the two solutions should match, as indeed they do. The close agreement between the numerical solution and the Fourier solution using only eleven terms suggests that eleven terms is sufficient to get accurate results from the Fourier method in this case. However, we could test this conclusion by including more than eleven terms to see if it results in any noticeable changes in the output.

4.7 The Pendulum

Although Sect. 4.1 shows that many physical systems can be approximated as harmonic oscillators, the analysis is restricted to oscillations near a stable equilibrium point. An important question remains: What happens if the motion deviates too far from the stable equilibrium point, such that we can no longer accurately approximate the system as a harmonic oscillator? This section looks at a simple, but important, case in which we can observe deviations from harmonic oscillator behavior when the amplitude of oscillations about a stable equilibrium point becomes too large.

Consider the simple pendulum, as depicted in Fig. 4.25: a point particle (or bob) of mass m attached to the end of a massless rigid rod of length L. The other end of the rod is fixed to a point, and the rod can pivot about this point to move in a single plane.

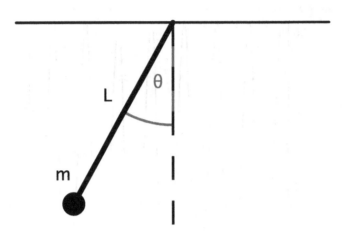

Fig. 4.25 The simple pendulum: a point mass attached to a massless, rigid rod that is free to rotate about the opposite end

Basic trigonometry implies that the pendulum bob lies a distance $L \cos \theta$ below the top of the pendulum. If we let the bottom of the pendulum serve as our zero-point for gravitational potential energy, then the potential energy function for this system is

$$U(\theta) = mgL(1 - \cos \theta). \tag{4.37}$$

The equilibrium points for this system occur at values of θ for which $dU/d\theta = 0$. To find the equilibrium points we first evaluate $dU/d\theta$.

```
(%i) Up(theta):=m*g*L*(1-cos(theta))$
     diff(Up(theta),theta);
(%o) g m sin (θ) L
```

The equilibrium points will occur whenever $\sin \theta = 0$, or when $\theta = n\pi$ for integer n. Although, mathematically, there are an infinite number of solutions, these solutions correspond to only two physically distinct positions of the pendulum: hanging straight down ($\theta = 0 \pm 2n\pi$) or sticking straight up ($\theta = \pi \pm 2n\pi$). We can evaluate the stability of these two equilibrium points by first computing $d^2U/d\theta^2$ and then evaluating this second derivative at each equilibrium point to find the corresponding value of k for that point.

```
(%i) k(theta):=''(diff(Up(theta),theta,2))$
     [k(0),k(%pi)];
(%o) [g m L, −g m L]
```

We find that at $\theta = 0$, $k = gmL > 0$ and therefore this equilibrium point is stable. However, at $\theta = \pi$, $k = -gmL < 0$ and therefore the equilibrium point at $\theta = \pi$ is unstable. These results are intuitive: a pendulum can balance if it is placed straight upward, but the slightest nudge will send it falling away from this upright position. In contrast, a downward hanging pendulum will only begin to oscillate if nudged away from its equilibrium position.

The potential energy function near the stable equilibrium point at $\theta = 0$ will have the approximate form of a harmonic oscillator potential energy function. We can illustrate this fact by expanding $U(\theta)$ as a Taylor series about $\theta = 0$.

```
(%i) taylor(Up(theta),theta,0,4);
```
$$(\%o) \quad \frac{gmL\theta^2}{2} - \frac{gmL\theta^4}{24} + \dots$$

To second order in θ we see that $U(\theta) \approx (1/2)gmL\theta^2$. We can compare this approximate function to the full potential energy function (Eq. 4.37) over the full range of motion for the pendulum ($-\pi \le \theta \le \pi$). We will assume a 1 m pendulum with a 1 kg bob. The code below generates a plot of the full potential energy function as well as the approximate function. The results are displayed in Fig. 4.26.

```
(%i) [g,m,L] : [9.8,1,1]$ wxdraw2d(key = "Actual",
    explicit(Up(theta),theta,-%pi,%pi),
    xlabel="{/Symbol q}", ylabel="U(J)",
    color=gray,key="Approximate",
    explicit(g*m*L*theta^2/2,theta,-%pi,%pi))$
```

Figure 4.26 shows that the approximate function (lighter curve) fits very well with the actual potential energy function (darker curve) near $\theta = 0$, but for $|\theta| > 1$ there are noticeable deviations among the curves on the scale of this plot. For large values of θ the two curves deviate significantly. So near the equilibrium point we

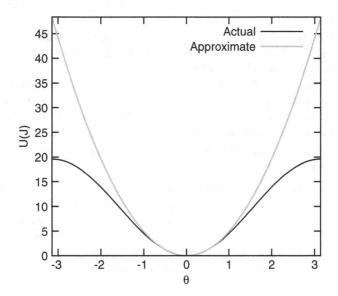

Fig. 4.26 Potential energy as a function of angle (in radians) for a pendulum near the stable equilibrium point (*black*) and second order Taylor series approximation (*gray*)

would expect the motion to mimic that of a harmonic oscillator, but far from the equilibrium point we expect significant deviations from harmonic oscillator motion.

In Sect. 4.2 we showed that the energy equation for a simple harmonic oscillator could be rewritten to demonstrate that the motion of a harmonic oscillator will lie along an ellipse in the phase space. We now try a similar procedure with the pendulum. The total energy for our simple pendulum is given by

$$E = (1/2)mv^2 + mgL(1 - \cos\theta) = (1/2)mL^2\omega^2 + mgL(1 - \cos\theta), \qquad (4.38)$$

where $\omega = \dot\theta$ is the angular velocity of the pendulum's motion. To determine the path of the pendulum in phase space we can solve Eq. 4.38 for ω as a function of θ.

```
(%i) kill(L,m,g)$ solve(E=m*L^2*%omega^2/2+
               m*g*L(1-cos(theta)),%omega);
```
$$(\%o) \quad [\omega = -\frac{\sqrt{2}\sqrt{\frac{E}{m}-gL(1-\cos(\theta))}}{L}, \omega = \frac{\sqrt{2}\sqrt{\frac{E}{m}-gL(1-\cos(\theta))}}{L}]$$

So the pendulum will follow a path in phase space that is defined by

$$\omega = \pm\frac{\sqrt{2}}{L}\sqrt{\frac{E}{m} - gL(1 - \cos\theta)}. \qquad (4.39)$$

The plus sign gives the portion of the curve for which the pendulum is swinging counterclockwise (positive angular velocity) while the minus sign gives the portion of the curve for which the pendulum is swinging clockwise (negative angular velocity). The code below constructs a plot of these phase space curves for three different energies: $E_1 = 5$ J, $E_2 = 15$ J, and $E_3 = 20$ J.[5] The resulting plot is shown in Fig. 4.27.

```
(%i) %omega(theta,E):=sqrt(2*E/m-2*g*L*(1-cos(theta)))
     /L$ L:1$ m:1$ g:9.8$ E1:5$ E2:15$ E3:20$
   wxdraw2d(yrange=[-2*%pi-1,10],
       /* For E = E1 */ key = concat("E =", string(E1)),
       explicit(%omega(theta,E1),theta,-%pi,%pi), key=
       "", explicit(-%omega(theta,E1),theta,-%pi,%pi),
       -Similar commands for E2 and E3 are omitted.-
       xlabel="{/Symbol q}",ylabel = "{/Symbol w}
       (rad/s)");
```

Figure 4.27 shows three different curves. Near the center is the ellipse-shaped curve for E_1. At this energy the pendulum has small amplitude oscillations, so it remains near the stable equilibrium point at $\theta = 0$. The phase space path looks just like that of a harmonic oscillator. Outside of this curve there is another closed curve for E_2. The curve is not quite ellipse-shaped, but has left and right sides that are somewhat pointed. At this energy the pendulum swings farther from equilibrium

---

[5]The Symbol commands might not work in all *Maxima* installations. The text entries theta and omega may be substituted.

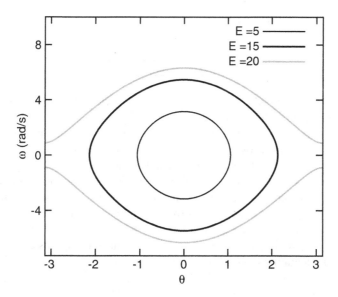

Fig. 4.27 Phase space path (angular velocity ω, in rad/s, as a function of angle θ, in rad) for a simple pendulum with three different energies

(reaching as far as two radians from the equilibrium point) and its motion deviates noticeably from that of a harmonic oscillator. However, the pendulum will still exhibit oscillatory motion.

For even greater energies, like E_3, the pendulum can swing with enough energy to reach the unstable equilibrium point at $\theta = \pm\pi$ and even go past it, so that the pendulum will repeatedly swing through full circles of motion. This motion is represented by the two curves at the top and bottom of the plot.[6]

How do these deviations from the elliptical phase space path affect other properties of a pendulum's motion? We have seen that one important property of a harmonic oscillator is that its motion is periodic with a period $T = 2\pi\sqrt{m/k}$. This period does not depend on the amplitude of the oscillations. Will the same hold true for the pendulum? If the pendulum has enough energy to swing all the way around, then it won't oscillate at all. But if the pendulum does oscillate, will the period of these oscillations depend on the amplitude of the motion?

We can determine the period of a pendulum by rewriting Eq. 4.39:

$$\omega = \frac{d\theta}{dt} = \pm\frac{\sqrt{2}}{L}\sqrt{\frac{E}{m} - gL(1 - \cos\theta)}. \qquad (4.40)$$

---

[6]Unless the option `draw_realpart=false` is used the graph will show horizontal lines at $\omega = 0$. This happens because, by default, `draw` plots the real part of complex-valued functions. We use the command `set_draw_defaults` at the beginning of the workbook to suppress the drawing of the real parts of complex values and well as to set other default values.

We can separate the variables θ and t to find

$$dt = \pm \frac{L\,d\theta}{\sqrt{2E/m - 2gL(1 - \cos\theta)}}. \tag{4.41}$$

We can integrate both sides of Eq. 4.41 to find the period of the pendulum, but we must first consider our limits of integration.

We want the time integral to cover one full period of oscillation. Therefore, the integral over θ must also cover a full oscillation. In other words, we need to integrate from one extreme value of θ to the other, and back again. Suppose the pendulum is released from rest at an angle $-\theta_{max}$. The pendulum will swing until it reaches the angle θ_{max}. At that point it will turn around and swing back. So our integral over θ must run from $-\theta_{max}$ to θ_{max} (with positive ω), and back again (with negative ω).

We can write the total energy E in terms of θ_{max} because when $\theta = \theta_{max}$ then $\omega = 0$, so Eq. 4.38 gives $E = mgL(1 - \cos\theta_{max})$. Substituting this expression into Eq. 4.41 and integrating both sides we find

$$\int_0^T dt = \int_{-\theta_{max}}^{\theta_{max}} \frac{L\,d\theta}{\sqrt{2gL(\cos\theta - \cos\theta_{max})}} + \int_{\theta_{max}}^{-\theta_{max}} \frac{-L\,d\theta}{\sqrt{2gL(\cos\theta - \cos\theta_{max})}}, \tag{4.42}$$

or

$$T = 2\int_{-\theta_{max}}^{\theta_{max}} \frac{L\,d\theta}{\sqrt{2gL(\cos\theta - \cos\theta_{max})}}. \tag{4.43}$$

We can try to evaluate this integral using *Maxima*'s `integrate` command.

```
(%i) 2*integrate(L/sqrt(2*g*L*(cos(theta) -
     cos(theta[max]))), theta,-theta[max],theta[max]);
(%o) 0.45175 ∫_{-θ max}^{θ max} \frac{1}{\sqrt{cos(θ)-cos(θ max)}} dθ
```

Maxima is unable to evaluate this integral symbolically. This is not a flaw in *Maxima*: this integral has no closed form solution. We can, however, compute this integral numerically as long as we have values for all of the parameters. Note that we can rewrite Eq. 4.43 as

$$T = 2\pi\sqrt{\frac{L}{g}}\frac{1}{\pi\sqrt{2}}\int_{-\theta_{max}}^{\theta_{max}} \frac{d\theta}{\sqrt{\cos\theta - \cos\theta_{max}}} = 2\pi\sqrt{\frac{L}{g}}f(\theta_{max}). \tag{4.44}$$

We use *Maxima*'s `quad_qags` numerical integration command to evaluate the $f(\theta_{max})$. Then we can construct a plot of $f(\theta_{max})$ versus θ_{max} to determine how the period of the pendulum's oscillations depend on amplitude. The code below shows how to construct the plot and Fig. 4.28 shows the result.

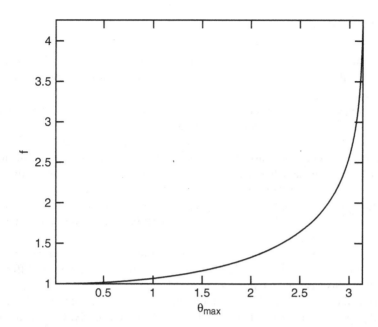

Fig. 4.28 Plot of the function $f(\theta_{max})$, where θ_{max} is the amplitude of the pendulum's oscillation (in radians). The period of a pendulum's oscillation is $T = 2\pi \sqrt{L/g} f(\theta_{max})$

```
(%i) f(q):=quad_qags(1/sqrt(cos(x)-cos(q)),x,-q,q)[1]
        /(%pi*sqrt(2))$
     wxdraw2d(explicit(f(q),q,0.01,%pi-0.01),xlabel=
        "{/Symbol q}_max",ylabel="f")$
```

Note how we use the [1] to select the first element from the list that quad_qags returns, so that the function returns the value of the integral but not the other numbers in that list. Note also how our plot range extends from just above 0 to just below π: the numerical integration routine runs into problems at $\theta_{max} = 0$ or π. The plot shows that for small values of θ_{max}, $f(\theta_{max}) \approx 1$. So for small amplitude oscillations the period of the pendulum is simply $T = 2\pi \sqrt{L/g}$ and the period does not depend on amplitude (since the curve is relatively flat). However, for larger amplitude oscillations the period will be longer. For $\theta_{max} = 2$ radians the period is roughly 50 % greater than it is for small amplitude oscillations. As θ_{max} approaches π the period increases without bound. This result makes sense because if the pendulum were released exactly at $\theta = \pi$ then it would remain at that unstable equilibrium point forever.

As the system moves far from the stable equilibrium point the motion begins to deviate from that of a harmonic oscillator. In the case of the simple pendulum the deviations are not dramatic. The pendulum still oscillates as long as it doesn't have enough energy to swing all the way around. In the next chapter, however, we will see that these deviations from a harmonic oscillator potential energy function can give rise to some novel and interesting behavior patterns.

4.8 Exercises

1. Show that if the quarter sphere is placed on its point (the point C in Fig. 4.2) then it still has an equilibrium at $\theta = 0$. Is this equilibrium stable or unstable? Provide proof for your answer.

2. Investigate the isotropic three-dimensional harmonic oscillator with $k_x = k_y = k_z = 1$ N/m. Plot the motion for several different initial conditions. Examine cases in which the oscillator starts from rest but is displaced from the origin (try several different directions) as well as cases where the oscillator is displaced from the origin and has a nonzero initial velocity. For what kind of initial conditions is the motion periodic? For what kind of initial conditions is the motion one-dimensional? For what kind of initial conditions is the motion two-dimensional? Is it possible for the motion to be three-dimensional (i.e., not confined to a plane)?

3. Repeat the previous problem, but this time for an anisotropic oscillator with k_x, k_y, and k_z equal to 1, $1/2$, and $1/3$ N/m, respectively. Comment on the differences between this case and the previous case.

4. Repeat the previous problem, but this time for an anisotropic oscillator with k_x, k_y, and k_z equal to 1, $1/\sqrt{2}$, and $1/\pi$ N/m, respectively. Comment on the differences between this case and the previous case.

5. Consider a damped harmonic oscillator with $\omega_0 = 1$ rad/s, $x_0 = 1$ m, and $v_0 = 0$. Construct a single plot that shows $x(t)$ for three different values of β: 0.5, 1, and 1.5 s^{-1}. Then do the same for plots of $v(t)$, phase space trajectory, and $E(t)$. For each plot, comment on the differences between the three cases. Make sure to clearly identify which case is underdamped, which case is overdamped, and which case is critically damped.

6. Examine what happens to the steady state motion of a driven harmonic oscillator, as shown in Fig. 4.16, if you vary some of the parameters. Recreate the figure using all of the same parameters except let $\beta = 0.5$ s^{-1}. How (if at all) does increasing β alter the period and amplitude of the steady state motion? Recreate the figure again, but this time just change f to 5 N/kg. How (if at all) does increasing f alter the period and amplitude of the steady state motion? Finally, recreate the figure with the same parameters except for the value of ω. Create plots using the following values of ω: 0.6, 0.8, 1.0, 1.2, and 1.4 rad/s. Discuss how these changes in the driving frequency alter the period and amplitude of the steady state motion.

7. Repeat the analysis of Sect. 4.6, but this time with a driving force given by

$$f(t) = f\omega \left(\mathrm{mod}\,(t + \pi/\omega, 2\pi/\omega) - \pi/\omega\right)/\pi. \qquad (4.45)$$

Note that the mod(x, y) function can be written in *Maxima* as mod (x,y). Plot the function, using $f = 1$ N/kg and $\omega = 1$ rad/s. Then investigate the motion of a harmonic oscillator with $\omega_0 = 1$ rad/s and $\beta = 0.5$ s^{-1} driven by this force. Use rk to generate a numerical solution for $x(t)$ if the oscillator starts from rest at the equilibrium position. Then plot the Fourier series for this $f(t)$ using the first ten terms in the series, and then again using the first 50 terms. Note how the inclusion of more terms leads to a better approximation of $f(t)$. Finally, plot the approximate steady state solution for $x(t)$ using the first ten terms in the Fourier series for $f(t)$.

8. A quartic oscillator has potential energy $U(x) = \alpha x^4$, where α is a positive constant. Show that this oscillator has an equilibrium point at $x = 0$. The stability of the equilibrium point is indeterminate using the methods discussed in this chapter, but it turns out that the equilibrium is stable. Plot the potential energy function (for some value of α) and explain how the plot indicates that the equilibrium point is stable. Find the period of oscillation for a particle of mass m in this quartic oscillator system if the particle is released from rest at $x = x_0$. (Note: you may get a cryptic answer from *Maxima* when you evaluate the integral needed to find the period, but just use float to convert the answer to a more useful form.) Does the period of a quartic oscillator depend on the amplitude of oscillation? How so?

Chapter 5
Physics and Computation

By this point we have seen that *Maxima*'s built-in routines can be very helpful tools in solving physics problems. Users can expand *Maxima*'s usefulness by taking advantage of its programming capabilities. Although *Maxima* should not be viewed as a substitute for a full-featured programming language, it does have some basic programming features that allow users to write simple programs. This chapter introduces these basic programming features and shows how they can be used to solve problems in mathematics and physics.

  Maxima's programming features not only let users add new functionality, but they also allow users to explore the algorithms used to carry out various numerical computations. Although many of these computational tasks can be performed using *Maxima*'s built-in routines, it is important for users to have some idea of what is going on "inside the black box." In this chapter we will explore some of these algorithms and show how the behavior of the algorithms can be connected to important physics concepts. Other numerical algorithms are discussed in the Appendix.

5.1 Programming: Loops and Decision Structures

Maxima is designed for symbolic and numerical mathematics, not as a tool for computer programming. Even so, *Maxima* does offer programming features that can be useful for mathematics and physics. This section introduces two types of programming features that will be used later in the book: *loops* and *decision structures*.

© Todd Keene Timberlake & J. Wilson Mixon, Jr. 2016
T.K. Timberlake, J.W. Mixon, *Classical Mechanics with Maxima*, Undergraduate
Lecture Notes in Physics, DOI 10.1007/978-1-4939-3207-8_5

5.1.1 Loops

Loops allow repetition. In each pass through a loop, *Maxima* executes a specific set of commands, possibly changing the values of certain variables. It then makes another pass through the loop, using the new values for the variables produced by the previous pass. *Maxima* uses the do command to specify the set of instructions to be executed in each pass of the loop. Other commands control how many times these instructions will be executed before the program stops. The best way to understand how loops work is to look at several examples. Our first example of a *Maxima* loop is shown below.

```
(%i) for i:0 thru 2 do (display(i))$
(%o) i = 0
     i = 1
     i = 2
```

The example above uses a for command to control the loop. The variable *i* is an *index* that counts the number of passes through the loop. This index variable is given an initial value of 0, and its value is automatically increased by 1 *after* each pass through the loop. The thru command specifies when *Maxima* should terminate the loop. In this case, we instruct *Maxima* to continue executing the loop until the value of *i* reaches 2. Then Maxima executes the loop one last time and stops. Finally, the do command tells Maxima that the next command after do is the one that should be executed on every pass through the loop. Each line of the output is produced by a different pass through the loop. The last pass occurs when $i = 2$. Note that the index *i* is a *local* variable, which means that its value is defined only within the loop, not stored in *Maxima*'s memory. Once *Maxima* exits the for loop the value of *i* reverts to whatever it was before the for loop was entered.

For this first example it may help to walk through a detailed description of what happens during each pass through the loop.

- On the first pass the variable *i* is set to 0 and the command display(i) is executed.
- Then the value of *i* is increased to 1 and the display command is executed again (giving a different output value this time, because the value of *i* has changed).
- Then *i* is increased to 2 and the display command is executed again.
- At this point the loop terminates because it has completed the pass with $i = 2$ as specified by the thru command.

The next example uses for in a different way. In this example the loop is controlled by the while command. This command specifies the condition under which the program will continue passing through the loop. In this case, the programming will continue looping as long as *i* is less than 4. As soon as the condition is no longer satisfied, then the loop will terminate.

```
(%i) for i:0 while i<4 do display(i)$
(%o) i = 0
     i = 1
     i = 2
     i = 3
```

It may seem that `thru` and `while` do the same thing, but there is at least one important difference. The `thru` command only lets you specify the final value for the index variable. The `while` command specifies a condition for continuing the loop that may involve variables other than the index variable. Look at the example below and try to predict the value of x after the program has been executed.[1]

```
(%i) x:4$ for i:0 while x<280 do x:x*x-3$
```

The next example uses the `unless` command to control the loop. This command is similar to the `while` command, except that it specifies a condition that causes the loop to terminate. In the example below the loop will continue as long as i is *not* greater than 3. Once i exceeds 3, then the loop terminates.

```
(%i) for i:0 unless i>3 do display(i)$
(%o) i = 0
     i = 1
     i = 2
     i = 3
```

To increase the value of the index variable by increments other than 1, use the `step` command. For example, the code shown below increases i by 3 after each pass through the loop, terminating after the $i = 9$ pass. Try changing `thru 9` to `thru 10` to see what happens. This example illustrates that the `thru N` command is equivalent to `unless i>N`, where i is the index variable.

```
(%i) for i:0 step 3 thru 9 do display(i)$
(%o) i = 0
     i = 3
     i = 6
     i = 9
```

We can change the value of the index variable in more complicated ways, using the `next` command. Look at the example shown below and be sure that you understand why it produces the output shown. For the first pass through the loop the index is set to 1. On the next pass the index is set to five times the current value of i, so the new value of i will be 5. On the next pass the new index will be the old index ($i = 5$) times five, so $i = 25$, and so on.

```
(%i) for i:1 next 5*i while i<1000 do display(i)$
(%o) i = 1
     i = 5
     i = 25
     i = 125
     i = 625
```

---

[1]To check your answer, have *Maxima* display the value of x immediately after executing the program.

We often want to execute more than one command during the course of a single pass through the loop. The `block` command lets us define a sequence of commands to be executed as a single step. The `block` command takes the form `block([v_1, ..., v_m], com_1, ..., com_n)` where `com_1, ..., com_n)` represent the sequence of commands that will be executed during each evaluation of the block. The commands will be executed in the order in which they are listed. The v_i are *local* variables within the block. The values may be changed within the block. Once *Maxima* exits from the block, however, these variables revert to whatever values they had before the block was executed. The examples below are two programs that compute the so-called Fibonacci numbers. The first uses a `block` to keep the created value(s) of the variables local; the second does not. Notice the different treatments of the z variable.

```
(%i) x:1$ y:1$ z:5$
     for i:3 thru 7 do
         block([z], z:x+y, x:y, y:z,print(y))$
       display(z)$
(%o) 2
     3
     5
     8
     13
     z = 5
(%i) x:1$ y:1$ z:5$
     thru 5 do
         (z:x+y, x:y, y:z,print(y))$
       display(z)$
(%o) 2
     3
     5
     8
     13
     z = 13
```

The following example uses a `block` to create the recursion required to compute the factorial of a number. For a given value of the variable x, the function of x equals x's current value times the value when x is one less than the current value. The lists show the factorials for the integers 1–5, first created by this small program and then created directly by using *Maxima*'s `factorial` command.

```
(%i) kill(values,functions,arrays)$
     g(x) := block([temp], temp: 1,
        while x > 1 do (
        temp : x*temp, x:x -1), temp )$
     xL : makelist(x, x, 1, 5);
     gL : map(g, xL);
     xfactorial_Maxima : map(factorial, xL);
(%o) [1, 2, 3, 4, 5]   (%o) [1, 2, 6, 24, 120]   (%o) [1, 2, 6, 24, 120]
```

5.1.2 Decision Structures

Maxima functions can be extended to include logical expressions like the following simple `if ...then ...else` statement. In this example, the result of a negative or zero argument for $f(x)$ is a text message warning that x must be positive. If the argument is positive then $f(x) = \log(x)$. The code below defines the function and generates output for the function evaluated at a few values of x.

```
(%i) f(x) := if x <= 0 then "x must be positive"
        else log(x)$
     f(-5.0); f(0); f(5.0);
(%o) x must be positive
(%o) x must be positive
(%o) 1.6094
```

The `if ...then` construction is called a *decision structure*, because it supplies the computer with alternatives and criteria to decide which alternative should be used. The `if ...then` structure can be extended to include intermediate possibilities, using `elseif` statements between the `if` and the `else`. The code below uses one such statement to create a piecewise function with kinks at $x = 1$ and at $x = 2$. The code also generates the plot of the function shown in Fig. 5.1.

```
(%i) f(x) := if x < 1 then x
        elseif x<2 then 1+3*(x-1)
        else 4-2*(x-2)$
     wxdraw2d(xlabel="x",ylabel="f(x)",
        explicit(f(x),x,0,4))$
```

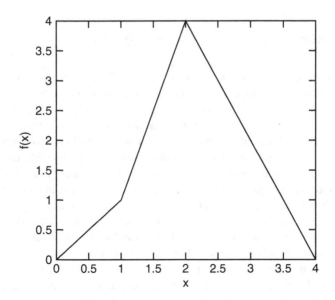

Fig. 5.1 Plot of a piecewise function constructed using an `if ... then` decision structure

Later in the book we will see that the combination of loops and decision structures can provide for powerful programming. To give an example, the code below defines a function that tests whether or not the input value is a prime number. The function uses *Maxima*'s `integerp` command, which evaluates whether or not the argument is an integer. It also uses *Maxima*'s `catch` and `throw` commands. For more information on these commands see the *Maxima* manual or *wxMaxima*'s Help menu. To illustrate the capabilities of this newly defined function, the code generates a table that shows the result of evaluating the function for several inputs.

```
(%i) prime(x):= (if integerp(x) then
     (if x<2 then "not an integer > 1" else
      catch(i:2,
        while i <= sqrt(x) do
         (j:2, unless j*i > x do
          (if j*i=x then throw("no")
          else j:j+1), i:i+1
          ),
        throw("yes")))
      else "not an integer")$
    /*Evaluate four numbers */
    transpose(matrix(["Number", 1, 17, 21, 21/2],
     ["Evaluation", prime(1), prime(17),
      prime(21), prime(21/2)]) );
```

$$
(\%o) \quad \begin{bmatrix} \text{Number} & \text{Evaluation} \\ 1 & \text{not an integer} > 1 \\ 17 & \text{yes} \\ 21 & \text{no} \\ \frac{21}{2} & \text{not an integer} \end{bmatrix}
$$

5.2 Random Numbers and Random Walks

Most of the material that this text covers is *deterministic*. That is, a specified set of conditions always generates the same outcome. Some processes, however, are not deterministic; they are *random*. Although classical physics does not involve the fundamental randomness of quantum mechanics, it is nevertheless true that many classical systems display motion that is apparently random. Moreover, random numbers can sometimes be used to produce completely predictable (and useful) results. This section briefly examines *Maxima*'s random number generation and how it can be used to investigate interesting mathematical and physical concepts.

Maxima's `random` command produces sequences of pseudorandom numbers. These numbers, actually the product of a mathematical algorithm that depends on a particular "seed," are generally close enough to random for practical purposes. The `random` command differs from most other *Maxima* commands in one important respect: simply using this command repeatedly can give different answers each time, depending on the seed. If we want to get the same "random" number sequence each time, we can begin with the commands below. These commands specify a seed, which sets the state of the random number generator.[2]

```
(%i) s1:make_random_state(345497)$
      set_random_state(s1)$
```

If x is a positive real (floating-point) number, then the command random (x) will generate a value selected from random real numbers that are uniformly distributed on the interval $[0, x)$. If x is a positive integer then random will generate a random integer from 0 to $x - 1$, with all possibilities being equally likely. Other inputs for x will generate an error message.[3]

```
(%i) /*floating point*/ random(1.0);
      /*integer*/ random(7);
(%o) 0.731    (%o) 5
```

Try evaluating the above commands a few times. If you reset the state of the random number generator to the state s1 before evaluating the random command, you always get the same result. If you evaluate the random command again without first resetting the state of the random number generator, then you can get a different result each time you evaluate random. Play around with this to make sure you understand how setting the state of the random number generator affects the results obtained from random.

5.2.1 Approximating π

As an example of producing a predictable outcome by using random numbers, consider the following method for approximating the value of π. Begin with a 2 by 2 square that is centered on the origin. Inscribe a unit circle, also centered on the origin. The square's area is 4; the circle's area is π. If we randomly generate a large number of points inside the square, then the fraction of these points that fall inside the circle should be $\pi/4$. That is, the *probability* that any one randomly generated value will fall within the unit circle is $\pi/4$. Therefore the number of randomly generated values (call this number j) that fall inside the unit circle approaches $n\pi/4$ as n becomes large. Another way to say this is that j tends toward $n\pi/4$, though it cannot equal $n\pi/4$ in any particular finite sample. This relationship provides an approximation, $\pi \approx 4j/n$, where j is the number of points that fall inside the unit circle and n is the number of randomly generated points inside the square.

---

[2]There is nothing special about the number that serves as the argument for the make_random_state command. Any integer may be used, but different arguments will generate different "seeds."

[3]Note the use of /* and */ to enclose comments in this code. Comments are ignored by *Maxima* when the code is evaluated.

The program below generates 10,000 pairs of x- and y-coordinates, each between -1 and 1. These values are generated by first generating a random number between 0 and 1, then subtracting 0.5, and finally multiplying by 2. For each coordinate pair, if $x^2 + y^2 < 1$, the point lies inside the square *and* inside the unit circle. When this condition is met, the counter variable j is increased by 1. Once all 10,000 points have been generated and tested, the program displays the approximate value of π given by $4j/n$. Suppose 7901 of the 10,000 points fall inside the circle, then the approximation of π is $4(7901/10{,}000) = 3.1604$. (But running the code again will give a different result, because it will generate a different set of random numbers, unless you reset the state of the random number generator each time.)

```
(%i) n:10000$ j:0$
     for i:1 thru n do
       block(x:2*(random(1.0) - 0.5),
         y:2*(random(1.0) - 0.5),
         if (x^2+y^2 <= 1) then j:j+1)$
     float(4*j/n);
(%o) 3.1764
```

The exercise above illustrates how the generation of random numbers can be used within a *Maxima* program, and how random numbers can be used to produce a predictable result. This method is not, however, an efficient way of calculating the value of π.

5.2.2 Evolution of an Ensemble

We can also use random numbers to illustrate the movement of an ensemble (or collection) of particles. For example, we can examine the motion of an ensemble of particles moving with constant acceleration. First we generate a large number of initial position and velocity values distributed uniformly within fixed ranges (i.e., within some rectangular region of the phase space). Think of these as initial conditions for a large number of particles. The code below shows how to generate these initial points and create a plot of the ensemble, which is shown in Fig. 5.2.

```
(%i) l1:makelist(
        [random(1.0),random(1.0)],i,1,4000)$
     wxdraw2d(xrange=[-0.1,1.1],yrange=[-0.1,1.1],
        xlabel="Initial Position (m)", ylabel="Initial
        Velocity (m/s)",points(l1))$
```

Now, allow each particle in the ensemble to move with constant acceleration a until some later time t. The code below evolves the ensemble to the time $t = 3$ s with constant acceleration $a = 9.8$ m/s^2. Figure 5.3 shows the final locations of the particles in phase space. The ensemble now occupies a distorted rectangle

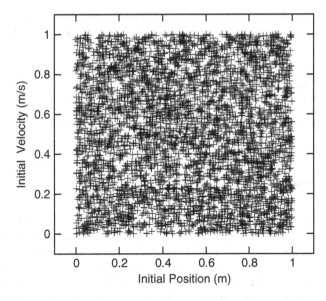

Fig. 5.2 Initial conditions for an ensemble of particles placed at random points within a rectangular region of phase space

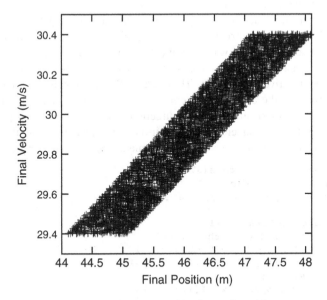

Fig. 5.3 Final phase space locations for the ensemble after motion with constant acceleration

(it's actually a parallelogram) in phase space. The fact that the area occupied by the ensemble remains constant illustrates Liouville's Theorem, which we discuss more fully in Sect. 5.4.

```
(%i) t:3$  f(x):=[x[1]+x[2]*t+0.5*9.8*t^2,x[2]+9.8*t]$
     12:map(f,11)$
     wxdraw2d(xrange=[44,48.1],yrange=[29.3,30.5],
        xlabel = "Final Position (m)",
        ylabel= "Final Velocity (m/s)",points(12))$
```

5.2.3 A Random Walk

In this section we use *Maxima* to generate and display a "random walk" on a lattice in two dimensions. Random walks can be used to model interesting physical processes like the diffusion of molecules in a gas. Let's look at a simple example of a random walk. The walker begins at the origin. During each time step the walker can move one unit in any of four directions ($\pm x$ or $\pm y$) with equal probability for each direction. The probability to increase x by one unit, which we can denote $p(x \to x+1)$, is 0.25. Likewise, $p(x \to x-1) = p(y \to y+1) = p(y \to y-1) = 0.25$. We can generate a random number between zero and one and then use the probabilities to determine in which direction the walker will move.

To investigate this random walk, we first generate the sequence of points in the walk. The code below executes a random walk by initializing x and y to zero and then, for each step in the walk, using a set of `if ... then ... elseif` statements to determine whether the walker moves in the positive x, negative x, positive y, or negative y direction. Think carefully about why we must use the various numbers in the condition statements (0.25, 0.5, 0.75) and the final `else` statement in order to ensure that each direction is assigned an equal probability. After each step has been taken the new position (x, y) is added (using the `append` command) to an array that stores the entire sequence of locations for the walker.

```
(%i) s1:make_random_state(373497)$
     set_random_state(s1)$
     x:0$ y:0$ rw1:[[x,y]]$ steps:20000$
     for i:0 thru steps do
        block(r1:random(1.0),
           if r1 < 0.25 then x:x+1
           elseif r1 < 0.5 then x:x-1
           elseif r1 < 0.75 then y:y+1
           else y:y-1,
           rw1:append(rw1,[[x,y]]) )$
```

To display the results of this random walk we create four different graphic objects, each of which displays the first N steps of the walk (with $N = 200, 2000, 10,000$, and 20,000). The code below shows how to create the graphic object for $N = 200$. Similar code is used to create the other three graphic objects, which are then displayed using the `wxdraw` command. Figure 5.4 shows the results.

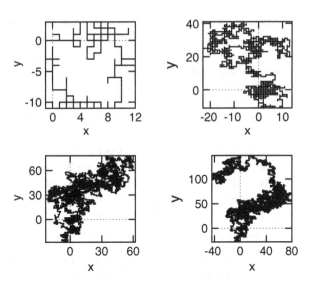

Fig. 5.4 The first N steps of a random walk for $N = 200$ (*top left*), $N = 2000$ (*top right*), $N = 10,000$ (*bottom left*), and $N = 20,000$ (*bottom right*)

```
(%i) n200:gr2d(xaxis=true, yaxis=true, xlabel="x",
        ylabel = "y", ytics=5, xtics=4,
        point_size=0, points_joined=true,
        points(makelist(rwl[i],i,1,200)))$

/*Commands for n2000, n10000, and n20000 omitted.*/

        wxdraw(n200,n2000, n10000, n20000,columns=2,
        dimensions=[480,480])$
```

We can also calculate the expected value of each coordinate after N steps. If the walker starts at (x_0, y_0) and can only move one unit in each of the four primary directions, then the expected value for x after N steps is:

$$\langle x \rangle_N = x_0 + N[p(x \rightarrow x + 1) - p(x \rightarrow x - 1)]. \tag{5.1}$$

Likewise the expected value for the y coordinate is

$$\langle y \rangle_N = y_0 + N[p(y \rightarrow y + 1) - p(y \rightarrow y - 1)]. \tag{5.2}$$

If the walker is not restricted to moving one unit at each step, then the expected value for x would be

$$\langle x \rangle_N = x_0 + N \sum_d [p(x \rightarrow x + d) \times d - p(x \rightarrow x - d) \times d], \tag{5.3}$$

where the sum is taken over all distances d that the walker can move in a single step. The expected value of y is obtained by replacing x with y in the above equations.

For the simple random walk we examined above the expected values for x and y are both zero, because the probability to increase each coordinate by one unit is balanced by an equal probability to decrease the coordinate by one unit. In fact, we can see that any time the probability for a coordinate to go up by d is equal to the probability for that coordinate to go down by d (for all d) then the expected value of that coordinate must be zero. But Fig. 5.4 shows that our random walker does not remain at the origin. The reason for this is that statistical fluctuations prevent the random sequence of steps from perfectly canceling, even though we would expect them to cancel in the limit of an infinite number of steps.

We examine these statistical fluctuations by looking at how the walker's distance from the origin changes as more steps are taken. The code below generates a list of these distances and then plots the distances as a function of the number of steps taken. The results are shown in Fig. 5.5.

```
(%i) rw2:makelist(sqrt(rw1[i][1]^2+rw1[i][2]^2),
        i,1,steps+2)$
     wxdraw2d(point_size=0, xlabel="Step Number",
       ylabel="Distance from Origin",
       points_joined=true,points(rw2),line_type=dots,
       explicit(sqrt(stepno),stepno,0,20000) )$
```

For a random walk of this type, in which the expected values of x and y are zero, we expect the distance from the origin, on average, to increase like the square root of the number of steps (shown by the dotted line in Fig. 5.5) because of statistical fluctuations. The plot shows that the distances are not generally equal to the square root of the step number, but the increase in distance with step number does generally track the curve for the square root function. For a random walk in which the expected values are not zero we would see a linear increase in distance as a function of the number of steps taken.

Fig. 5.5 Distance from the origin as a function of the number of steps for the random walk shown in Fig. 5.4

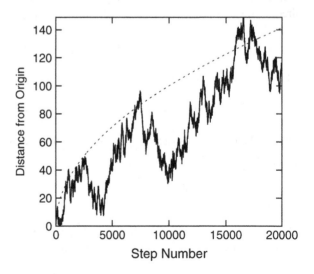

5.2.4   Nonuniform Distributions

We have so far been considering only random numbers that are uniformly distributed on some interval (and random integers). In this section we look at how to generate random real numbers that follow a nonuniform distribution. For comparison, the first example shows random values drawn from the uniform distribution on the interval (2,5). The `makelist` command generates the list of random numbers, while the `wxhistogram` command (part of the `descriptive` package) generates a histogram of the data with 20 bins, as shown in Fig. 5.6.

```
(%i) load(descriptive)$
     uniform:makelist(2+random(3.0),i,1,1000)$
     wxhistogram(uniform,nclasses=20, xlabel="x",
       ylabel="Frequency")$
```

The histogram is relatively flat, as we would expect for a uniform distribution. The numbers produced by the `random` command are equally likely to be found at any location in the interval $(2, 5)$. However, the histogram is not perfectly flat because our list of random numbers will exhibit some random fluctuations away from the expected uniform behavior.

The next example shows how to generate random numbers that follow a exponential distribution such that the probability of getting a value between x and $x + dx$ is $P(x)dx$ where $P(x) = \exp(-x)$. It is fairly easy to derive that we need: $x = -\log(y)$ where y is distributed uniformly on $(0, 1)$. The code below uses this result to generate a list of 1000 random numbers that follow the exponential

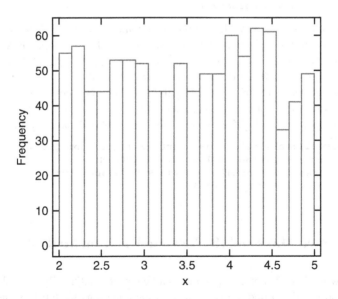

Fig. 5.6 Histogram for a set of 1000 random numbers uniformly distributed on the interval $(2, 5)$

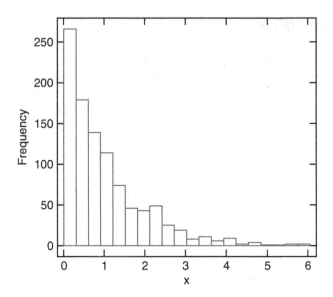

Fig. 5.7 Histogram of 1000 random numbers that follow the exponential distribution

distribution and then plot a histogram of those numbers. The histogram in Fig. 5.7 shows the expected decreasing exponential form.

```
(%i) expdist:makelist(-log(random(1.0)),i,1,1000)$
      wxhistogram( expdist,nclasses=20, xlabel="x",
      ylabel="Frequency")$
```

The final example shows how to generate numbers that follow a Gaussian (normal) distribution with mean zero and variance 1. This example uses the Box-Muller method where r is a Poisson random number and θ is a uniform random angle on the interval $(0, 2\pi)$. Then $x = \sqrt{2r}\cos\theta$ and $y = \sqrt{2r}\sin\theta$ are both distributed according to the normal distribution. The code below generates 1000 random numbers that follow the Gaussian (normal) distribution. The histogram of these numbers shown in Fig. 5.8 exhibits the expected "bell curve" form.

```
(%i) normal1:makelist( float(sqrt(-2*log(random(1.0)))*
      cos(2*%pi*random(1.0))),i,1,1000)$
      wxhistogram(normal1,nclasses=20, xlabel="x",
      ylabel="Frequency")$
```

5.3 Iterated Maps and the Newton–Raphson Method

As we saw in Chap. 1, not all equations can be solved analytically using *Maxima*'s solve command. To find the solution to some equations we must use numerical methods. In this section we will examine one particular method for solving

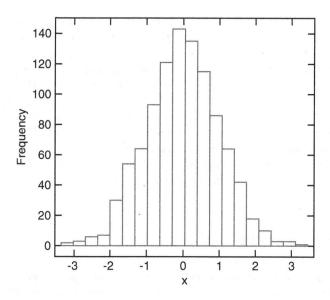

Fig. 5.8 Histogram of 1000 random numbers that follow a Gaussian (normal) distribution with mean 0 and variance 1

equations. This method is known as the Newton–Raphson method (or sometimes just Newton's method). A special case of the method was employed by Isaac Newton and simplified by Joseph Raphson, both in the seventeenth century. The general method, as described in this section, was first presented by Thomas Simpson in the eighteenth century.

The detailed examination of the Newton–Raphson method presented in this section serves two purposes. First, it gives you a "behind the scenes" look at how numerical root finding works. While the main goal of this book is to help you use *Maxima* as a tool for solving physics problems, it is good to know *some* details of how *Maxima* actually carries out all of these calculations. All of the *Maxima* routines that carry out numerical computations use some sort of *algorithm*, or set of specific operations. All algorithms have their strengths and weaknesses, and it is important to understand that fact even when you don't have to program the algorithm yourself.

The second purpose for examining the Newton–Raphson method is that it provides an introduction to the topic of *iterated functions*. To iterate a function means to apply the same function over and over again. Start with some initial value and substitute this value into the function. Take the result (the output of the function on this pass) and substitute that value back into the same function. Continue this process as many times as desired. This process can be encapsulated in a formula:

$$x_{n+1} = g(x_n). \qquad (5.4)$$

Here $g(x)$ is the function to be iterated. The initial value is x_0. Iterating the function using Eq. 5.4 generates a sequence of values $x_1, x_2, \ldots$.

We can think of an iterated function (also called an *iterated map*) as generating the dynamics of a very simple system. The variable x represents the state of the system. The initial state (at $t = 0$) is $x = x_0$. We generate future states of the system (x_1, x_2, etc.) by iterating the function $g(x)$. The sequence of x values produced by this iterated function forms a "trajectory" in the (one-dimensional) state space.

In Chap. 6 we will take a closer look at the dynamics of iterated maps to discover that these very simple dynamical systems can exhibit a rich variety of behaviors. In this chapter we will only examine the behavior of the iterated function used in the Newton–Raphson method. This method defines a function $g(x)$ that, if iterated, will converge to a root of the function $f(x)$. To get an idea of how this convergence works, let's take a look at a particular example of an iterated function.

5.3.1 Iterated Functions and Attractors

To better understand how an iterated function can converge to a particular value, we consider a specific example. We can then use *Maxima* to generate the sequence produced by iterating that function. Let's consider the function

$$g(x) = \sin(2x). \tag{5.5}$$

What happens if we iterate this function, starting with the initial condition $x_0 = 0.5$? The code below generates the first five iterations of the function ($x_1, \ldots, x_5$).

```
(%i) x0:0.5$      g(x):= sin(2*x)$
        for i:0 thru 4 do
           (x0:g(x0), print(x0))$
(%o) 0.84147
        0.99372
        0.91445
        0.96687
        0.93485
```

As an exercise, modify the code above to generate the first 30 iterations of the function $g(x) = \sin(2x)$ with initial condition $x_0 = 0.5$. You should find that the sequence converges to the value $x_* \approx 0.94775$. This value is an example of an *attractor*. After repeated iterations of the function, the sequence converges to this particular value and remains there. An attractor like this one is known as a *fixed point* because the tail end of the sequence just repeats a single value. The value to which this sequence converges must be a solution to the equation $g(x_*) = x_*$, so that substituting x_* into the function g just gives back x_* again. You can verify that $x_* \approx 0.94775$ is, indeed, a solution to $\sin(2x) = x$.

There are other kinds of attractors in which the sequence converges to a repetitive sequence of two, three, four, etc. numbers. We do not consider these here because they are not relevant to the Newton–Raphson algorithm. We will, however, see these other types of attractors in Chap. 6 when we examine chaotic dynamics.

So our function $g(x) = \sin(2x)$ has an attracting fixed point at $x_* = 0.94775$. Is that the only fixed point for this function? It is not hard to show that $x = 0$ is also a fixed point for $g(x)$. Why, then, doesn't our iterated function sequence converge to $x = 0$ instead of $x = x_*$? It seems as though some fixed points are attractors and others are not. To understand the difference between attracting and non-attracting fixed points, we analyze the *stability* of each fixed point.

A fixed point of an iterated function can be stable or unstable, in much the same way that the equilibrium points discussed in Sect. 4.1 can be stable or unstable. If the fixed point is stable then nearby points will tend to get closer to the fixed point when the function is iterated. If the fixed point is unstable then nearby points will tend to move away from the fixed point as the function is iterated. We can examine the stability of a fixed point by expanding the function in a Taylor series about the fixed point. If the function has a fixed point at $x = x_*$ then

$$g(x_* + \Delta x) \approx g(x_*) + \Delta x \left(\frac{dg}{dx}\right)_{x=x_*} = x_* + \Delta x \left(\frac{dg}{dx}\right)_{x=x_*}. \tag{5.6}$$

If our initial x value is a distance $|\Delta x|$ away from x_*, then after one iteration of the function it will be a distance

$$|g(x_* + \Delta x) - x_*| = \left|\Delta x \left(\frac{dg}{dx}\right)_{x=x_*}\right| \tag{5.7}$$

from the fixed point. So we see that if

$$\left|\left(\frac{dg}{dx}\right)_{x=x_*}\right| < 1 \tag{5.8}$$

then iterating the function brings the value closer to the fixed point and the fixed point is stable. On the other hand, if

$$\left|\left(\frac{dg}{dx}\right)_{x=x_*}\right| > 1 \tag{5.9}$$

then iterating the function takes the value farther away from the fixed point and the fixed point is unstable. The case

$$\left|\left(\frac{dg}{dx}\right)_{x=x_*}\right| = 1 \tag{5.10}$$

requires further analysis and will not be considered here.

Applying these criteria to the fixed points of $g(x) = \sin(2x)$ we find that the fixed point near $x = 0.94775$ is stable while the fixed point at $x = 0$ is unstable. The stability of a fixed point can also be illustrated by choosing an initial x value that is very close to the fixed point and observing the sequence of terms generated by iterating the function. If the terms get farther away from the fixed point, then the fixed point must be unstable. If they get even closer to the fixed point, then the fixed point is stable.

You might wonder whether an iterated function sequence always converges to a stable fixed point. It turns out that it does not always do so. As we have mentioned, a given function can have more than one stable fixed point, and obviously a single initial condition cannot converge to multiple stable fixed points. But even if there is only one stable fixed point, we cannot guarantee that our iterated function will converge to that fixed point for all initial conditions. Recall that our stability analysis was based on a first-order Taylor series expansion of the function near the fixed point. This analysis is reliable only in the region near the fixed point, where the first-order Taylor series approximates the function well. Outside of that region the iterated function may behave in a way that disagrees with the stability analysis given above.

What do we mean by "near" the fixed point? The answer is sensitive to the details of the function. Some stable fixed points attract a wide range of other points, while other stable fixed points attract only those points that are very close by.

The set of points that is attracted to a stable fixed point under iteration of the function is known as the *basin of attraction* of that fixed point. The basin of attraction may be large or small, and it can even consist of disjoint sets of points (i.e., points that are not all next to each other). The basin of attraction can even have a complicated *fractal* structure.[4] Here we will not attempt to determine the basin of attraction for any fixed points, but it is important to be aware that only points within the basin of attraction will converge to the fixed point under iteration of the function. This is particularly important when a function has multiple stable fixed points, which may have basins of attraction that intermix with each other.

5.3.2   The Newton–Raphson Method

Now that we have discussed the convergence of iterated functions, we can examine the details of the Newton–Raphson method. To use the Newton–Raphson method, we first rewrite the equation we want to solve so that all terms are on the left side. Now solving our equation becomes equivalent to finding the roots of some function $f(x)$. The roots of the function are just those values of x such that $f(x) = 0$. A given function may have no roots, or it may have several (or even an infinite number). As we will see, there is no completely reliable way to find all of the roots of a function. However, if we can make a good guess about the approximate location of the root, then the Newton–Raphson method is very likely to lead to a correct value (to whatever precision we desire) for the root.

---

[4] We will examine some fractal structures in Chap. 6.

The Newton–Raphson algorithm uses an iterated function $g(x)$ to generate a sequence of values that has a stable fixed point at a root of the function $f(x)$. As long as we start within the basin of attraction of this stable fixed point, then continued iterations of $g(x)$ should bring us arbitrarily close to the desired root of $f(x)$. To see how this works, suppose our initial guess for the value of the root is $x = x_0$. We expand $f(x)$ in a first-order Taylor series about $x = x_0$:

$$f(x) \approx f(x_0) + (x - x_0)\left(\frac{df}{dx}\right)_{x=x_0} = f(x_0) + (x - x_0)f'(x_0). \tag{5.11}$$

The right side of Eq. 5.11 provides an equation for a straight line, passing through the point $(x_0, f(x_0))$ and tangent to $f(x)$ at that point. This straight line serves as an approximation to the function, valid near $x = x_0$. We want to find the root of our function, but we may be able to get an approximate answer by determining where this straight line approximation crosses the x-axis. Setting the right side of Eq. 5.11 equal to zero and solving for x yields

$$x = x_0 - \frac{f(x_0)}{f'(x_0)}. \tag{5.12}$$

If the function $f(x)$ is linear, we are finished! The value of x given by Eq. 5.12 is a root (in fact, the one and only root) of $f(x)$. You can easily verify this for $f(x) = mx + b$, where m and b are constants. Even if the function is not linear, the value of x given by Eq. 5.12 will be closer to the root of $f(x)$ than x_0 was, provided x_0 was close enough to the root to begin with.

This is easier to state in the language of iterated functions. Define the function

$$g(x) = x - \frac{f(x)}{f'(x)}. \tag{5.13}$$

We can show that the function $g(x)$ has a fixed point at $x = x_*$ where $f(x_*) = 0$. In other words, the fixed point of $g(x)$ is a root of $f(x)$. Furthermore, we can show that this fixed point is stable. So if x_0 is within the basin of attraction of x_*, then $g(x_0)$ will be closer to x_* than x_0 is. By continuing to iterate the function $g(x)$ we get closer and closer to x_*. In principal we can get as close as we want, provided we iterate the function enough times. This is how the Newton–Raphson method works: to find a root x_* of $f(x)$ construct the function $g(x)$ and iterate it, starting with a value x_0 which is in the basin of attraction of the stable fixed point x_* of $g(x)$. Eventually this iteration process produces a value that is sufficiently close to the root x_*.

We must define what we mean by "sufficiently" close to the root, so that we have a criterion for stopping the iteration process. The key is that as our iterated function sequence approaches the desired root, the differences between successive values in the sequence decrease exponentially. One simple way to estimate how far we are from the root is to use the difference between the last two values in the sequence.

Once this difference is smaller than the specified precision for the root, we can stop iterating the function and use the final value as our numerically determined root.

For the Newton–Raphson method to work, we *must* choose an initial value x_0 that is in the basin of attraction of the fixed point x_* (under iteration of $g(x)$). If x_0 is not in the basin of attraction then the Newton–Raphson method may not converge at all (in which case the algorithm will never stop) or else it may converge to a different root than the one we seek. There is no easy way to determine the basin of attraction for a given root of an arbitrary function, so we must watch out for the bad behavior described above (either non-convergence or convergence to the wrong root). If we detect this bad behavior we can start over again with (we hope) a better value of x_0.

How can we determine what initial value to use? We want to select a value for x_0 that is close to the actual root. The easiest way to determine the approximate location of a root is to plot the function and see where the plot crosses the horizontal axis. We will illustrate this process, and the Newton–Raphson method, with the example function

$$f(x) = \tan(x) - x - 5. \tag{5.14}$$

We can plot this function to find approximate values for the function's roots. In order to locate the roots of a function we may need to experiment with the range of x values to use in the plot. We want to find a range of values that includes the root we seek, but which also makes the approximate location of that root easy to see. The code below defines our function and generates a plot that will allow us to determine approximate values for two of the function's roots. The plot is shown in Fig. 5.9.

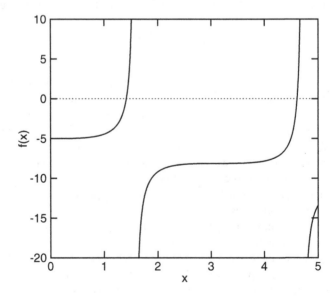

Fig. 5.9 Plot of the function $f(x) = \tan(x) - x - 5$

```
(%i) f(x) := tan(x) - x- 5$
     wxdraw2d(yrange=[-20,10],xaxis=true,line_width=2,
        xlabel="x",ylabel="f(x)",explicit(f(x),x,0,5))$
```

The curve crosses the x-axis near $x = 1.5$ and again near $x = 4.5$. Extending the range of this plot would reveal an infinite number of roots for this function. This result follows from the periodic nature of the tangent function. We will focus on finding the first root, near $x = 1.5$, using the Newton–Raphson method. The method can then be used to find any other root of this function, if desired.

Now that we know our function has a root near $x = 1.5$ we can use this value as the initial condition for our Newton–Raphson algorithm. The code below illustrates how to implement the algorithm to find the root of $f(x)$ that is near $x = 1.5$. First, we define a new function that is the derivative of $f(x)$, and then we define the function $g(x)$ that is used in the Newton–Raphson algorithm. Then the code specifies an initial value x0 as well as an error tolerance tol.

The for loop will run until the difference between consecutive terms in the iterated function sequence is smaller than the error tolerance, at which point the program exits the loop. On each pass through the loop the code outputs the number of iterations that have been performed and the current estimate for the root. Note that the code includes a safeguard that stops the loop after 30 iterations, in case the sequence does not converge.

```
(%i) df(x):="(diff(f(x),x));     (%o) df(x) := sec(x)² − 1
(%i) g(x):= x - f(x)/df(x)$      g(x);
(%o) x − tan(x)−x−5
           ────────────
           sec(x)²−1
(%i) x0:1.5$ tol:0.0001$ x1: g(x0)$ n:0$
     for i:1
       while ((i<30) and (abs(x1-x0) > tol)) do
         (x0:x1, x1:g(x1), n:n+1,
           print( n, x1))$
(%o) 1  1.4297
     2  1.4174
     3  1.4162
     4  1.4162
```

The algorithm converges after only four iterations. If we demand more precision by reducing the error tolerance to 10^{-6}, the Newton–Raphson method takes five iterations to converge to an answer (you can verify this by modifying the code— you may also want to increase the value of fpprintprec in order to display more decimal places in the results). What if we start with a different initial value? Try an initial value of 0.2. The iterations head off toward positive infinity, stopping only when the safeguard kicks in after 29 iterations. This initial value is not in the basin of attraction for the root at $x \approx 1.416$, so the Newton–Raphson method fails to converge to the desired root (or, in fact, to any root at all). Now try an initial value of 4.6 and confirm that the method does converge, but to a different root of $f(x)$.

These experiments illustrate two important facts about the Newton–Raphson method. First, when the method works it works very well. If the initial value is in the basin of attraction for a root, the method quickly converges to that root in only a few iterations even for relatively strict error tolerances. Second, the Newton–Raphson method does not always work. If the initial value is not in the basin of attraction for the desired root, then the method may converge to a different root or may fail to converge at all. Because it is hard to know in advance whether our initial value is in the basin of attraction, the Newton–Raphson method is not always the best choice for numerical root finding.

Maxima has a built-in routine called `newton` for implementing the Newton–Raphson method. To use the `newton` command we must first load the `newton1` package. The arguments of the `newton` command are the function, the variable for which we are solving, an initial guess for the root, and an error tolerance. If we apply this command to our example function we find that the built-in routine produces exactly the same result we found above.

```
(%i) load(newton1)$     newton(f(x),x,1.5,0.0001);
(%o) 1.4162
```

We examine another common method for root finding, known as the *bisection* method, in Sect. A.1. The bisection method is very reliable (in the sense of always finding the desired root) but it is also much slower than the Newton–Raphson method. *Maxima*'s built-in routine `find_root`, which we introduced in Chap. 1, uses a combination of the bisection method and linear interpolation (similar to the Newton–Raphson method, but without the need to know the derivative of the function) to achieve both reliability and speed. We can verify that `find_root` gives us the same answer as the Newton–Raphson method for our example function. Recall that `find_root` requires the user to specify bounds for x that contain the desired root of the function (and no other roots).

```
(%i) find_root(f(x),x,1.4,1.5);     (%o) 1.4162
```

5.4 Liouville's Theorem and Ordinary Differential Equation Solvers

In Chap. 2 we saw that *Maxima* has a routine (`rk`) for numerically solving systems of ordinary differential equations (ODEs). As with the other numerical computation routines in *Maxima*, the `rk` routine uses a specific algorithm to generate the solution to the system of ODEs given some initial conditions. The Runge–Kutta algorithm that is used for the `rk` command is presented in Sect. A.3. In this section we examine two simpler algorithms for numerically solving systems of ODEs: the Euler algorithm, and the closely related Euler–Cromer algorithm.

As with the Newton–Raphson method discussed in the previous section, one reason to learn about ODE solving algorithms is just to know what is going on "behind the scenes" with commands like `rk`. But the Euler and Euler–Cromer algorithms also provide a valuable opportunity to learn about an important physics topic known as *Liouville's Theorem*, which shows a connection between conservation of energy and preservation of area (or volume) in phase space. In this section we introduce these two ODE solving algorithms and show how their behavior illustrates Liouville's Theorem.

To investigate ODE solving algorithms we must, of course, first have a system of ODEs that we want to solve. The most commonly encountered ODE in Classical Mechanics is Newton's Second Law. For a single particle in one spatial dimension (with coordinate x) we can write Newton's Second Law as

$$\frac{d^2x}{dt^2} = \frac{F}{m}, \tag{5.15}$$

where F is the force on the particle and m is the particle's mass.

Most algorithms for solving ODEs are designed to work with first-order differential equations. Newton's Second Law is, however, a second-order differential equation. Fortunately, it is easy to reduce Newton's Second Law to two first-order equations by introducing the velocity, $v = dx/dt$. This definition, along with Newton's Second Law, gives us the following two first-order ODEs:

$$\frac{dx}{dt} = v, \tag{5.16}$$

$$\frac{dv}{dt} = \frac{F}{m}. \tag{5.17}$$

The material that follows focuses on this system of ODEs. Note that, in general, the force F can be a function x, v, and t.

5.4.1   The Euler Algorithm

The simplest numerical algorithm for solving the system of ODEs in Eq. 5.16 is the Euler algorithm. The Euler algorithm approximates the values of x and v at time $t + \Delta t$ using the values at time t. This means that we calculate the values of x and v only at a set of discrete times given by

$$t_n = t_0 + n\Delta t, \tag{5.18}$$

where n is a nonnegative integer. The values of x and v at these discrete times are denoted by $x_n = x(t_n)$ and $v_n = v(t_n)$.

To approximate x_{n+1} and v_{n+1} from x_n and v_n we can just try to use a first-order Taylor series expansion for $x(t)$ and $v(t)$:

$$x(t_n + \Delta t) \approx x(t_n) + \left(\frac{dx}{dt}\right)_{t=t_n} \Delta t,$$

$$v(t_n + \Delta t) \approx v(t_n) + \left(\frac{dv}{dt}\right)_{t=t_n} \Delta t. \tag{5.19}$$

Using Eq. 5.16 and our discrete notation we can rewrite Eq. 5.19 as

$$x_{n+1} \approx x_n + v_n \Delta t,$$

$$v_{n+1} \approx v_n + \left(\frac{F_n}{m}\right) \Delta t, \tag{5.20}$$

where $F_n = F(x_n, v_n, t_n)$ is just the force function evaluated at $t = t_n$. Equation 5.20 is known as the Euler algorithm. Once we specify the initial position and velocity $(x_0$ and $v_0)$ we can use this algorithm to generate the position and velocity at all later times t_n. Note that this is just like a two-dimensional version of the iterated maps we examined in the previous section.

It helps to consider a concrete example of using the Euler algorithm. Doing so is the best way to understand how the algorithm works, and it also reveals some problems inherent in this simple algorithm. Consider a simple harmonic oscillator subject to a Hooke's Law force:

$$F(x) = -kx.$$

We found the exact solution for this system in Sect. 4.2, but here we will examine how to generate an approximate solution for this system using the Euler algorithm. The Euler algorithm for this system is:

$$x_{n+1} = x_n + v_n \Delta t, \tag{5.21}$$

$$v_{n+1} = v_n - kx_n \Delta t / m. \tag{5.22}$$

Below we implement this algorithm using *Maxima*'s for command. First we define the values of our parameters ($k = 1N/m$, $m = 1kg$, $\Delta t = 0.2s$, and $nt = 100$ is the number of time steps to use). Then we create arrays of length nt to hold x and v values (we name these arrays xe and ve to indicate that they are computed using the Euler algorithm). Then we define the initial conditions ($x(0) = 1m$, and $v(0) = 0$). The for loop then applies the Euler algorithm $nt - 1$ times.

```
(%i) k:1$ m:1$ dt:0.2$ nt:100$
     array(xe,nt)$
     array(ve,nt)$
     xe[0]:1$ ve[0]:0$
     for i:0 while i < nt do
        block(xe[i+1]:xe[i]+ve[i]*dt,
        ve[i+1]:ve[i]-k*xe[i]*dt/m)$
```

Each pass through the for loop generates the values of *xe* and *ve* at the next time step. Immediately after the *for* statement we initialize an index that counts the number of passes through the loop ($i : 0$). This index is automatically increased by one after each pass through the loop. Then comes a statement which tests to see whether or not a sufficient number of passes have been performed (while i < nt). This is followed by the do command and then a block statement. The block statement

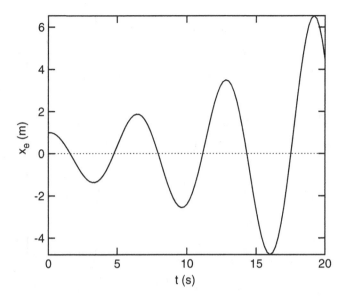

Fig. 5.10 Numerical solution for $x(t)$ in the harmonic oscillator system generated by the Euler algorithm

gathers all the code for the algorithm so that it can be executed as a single unit. Once the for loop is executed it has produced values for x_n and v_n for $n = 1$ to $nt - 1$.

Next we plot the results using wxdraw2d. The code below produces a plot of x as a function of t, and Fig. 5.10 shows the result. The makelist command first produces a list of ordered pairs of the form (t, x). The wxdraw2d command generates a plot of these points. The points_joined=true option specifies that lines should be drawn to "connect the dots." We could redo the plot with points_joined=false to see the sequence of points generated by the algorithm.

```
(%i) xvst:makelist([i*dt,xe[i]],i,0,nt)$
        wxdraw2d(xlabel="t (s)",ylabel="x_e (m)",xaxis=true,
        point_size=0,points_joined=true,points(xvst))$
```

We can compare this result to the exact solution found in Sect. 4.2 by plotting both the Euler algorithm solution (in black) and the exact solution ($x(t) = \cos(t)$, shown by the thicker gray curve) using the code below. The resulting plot is shown in Fig. 5.11.

```
(%i) wxdraw2d(xaxis=true,user_preamble="set key top
        left", point_size=0,points_joined=true,
        xlabel="t (s)",ylabel="x (m)",key="Euler x_e(t)",
        points(xvst),color="light-gray",key="Exact x(t)",
        line_width = 3,explicit(cos(t),t,0,nt*dt))$
```

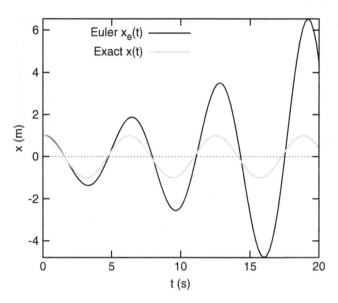

Fig. 5.11 Euler approximation solution and exact solution for $x(t)$ for the simple harmonic oscillator

Clearly there is a problem with our Euler algorithm result. We discuss this problem below, but first we look at the plot, in Fig. 5.12, of the velocity versus time (for the Euler algorithm as well as the exact solution $v(t) = -\sin(t)$), constructed in a similar way.[5]

Plotting the solution in phase space (v versus x) is instructive. The code below constructs a plot of v versus x for the range of times we have computed and compares this approximate solution to the exact path in phase space. The resulting plot is shown in Fig. 5.13.

```
(%i) xvsv:makelist([xe[i],ve[i]],i,0,nt)$
     wxdraw2d( user_preamble = "set size ratio 1",
        xaxis = true, yaxis=true,point_size=0,
        point_type=6,xlabel="x (m)", ylabel="v (m/s)",
        points_joined=true, points(xvsv),
        line_width=3,color=gray,
        parametric(cos(t),-sin(t),t,0,2*%pi))$
```

The exact solution produces an ellipse in the phase space. In contrast, the values generated by the Euler algorithm spiral outward. We now construct a plot of the total

---

[5]The commands are essentially identical to those above and are omitted.

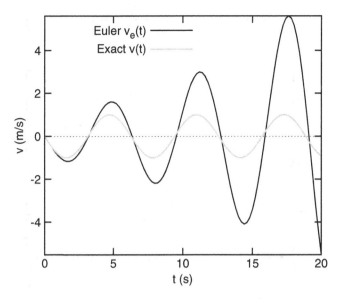

Fig. 5.12 Euler approximation and exact solution for $v(t)$ for the simple harmonic oscillator

Fig. 5.13 Euler approximation (*thin black curve*) and exact phase space path (*thick gray curve*) for the simple harmonic oscillator

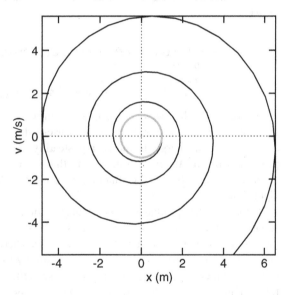

energy of the oscillator as a function of time, using our Euler algorithm solution. The total energy is given by

$$E = \frac{1}{2}mv^2 + \frac{1}{2}kx^2. \tag{5.23}$$

The code below constructs a list of ordered pairs of (t, E) and then plots these points. The resulting plot is shown in Fig. 5.14.

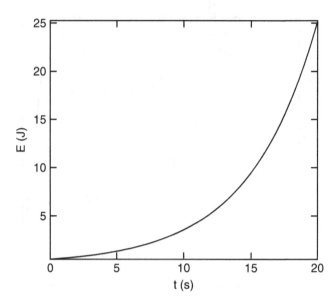

Fig. 5.14 Euler approximation for the total energy as a function of time for the simple harmonic oscillator

```
(%i) evst:makelist([i*dt,0.5*k*xe[i]^2+0.5*m*ve[i]^2],
         i,0,nt)$
     wxdraw2d(point_size=0,points_joined=true,
         xlabel="t (s)", ylabel="E (J)",points(evst))$
```

Figure 5.14 illustrates a major problem with this application of the Euler algorithm: energy is supposed to be conserved in this system but in the Euler algorithm result energy is increasing steadily. This anomaly relates to another problem with the Euler algorithm, one that we can illustrate by looking at how an ensemble of orbits (rather than a single orbit) behaves under the action of the algorithm.

The code below produces 1000 initial conditions located at random inside a small square around $x = 1$, $v = 0$. It then uses the Euler algorithm to calculate where these trajectories will be at $t = 20$ s. Note that there are three for loops in this code. The first is fairly straightforward and simply loops over the 1000 particles in order to set their initial conditions, making use of the random command to generate random initial values. The next for loop iterates the Euler algorithm for $nt-1$ steps, but this loop contains another for loop within it. This procedure of putting one loop inside another is known as "nesting" loops. Here the "outer" loop cycles over the time steps of the Euler algorithm, while the "inner" loop cycles through all of the different particles. Spend some time thinking through this code to determine exactly what it is doing.

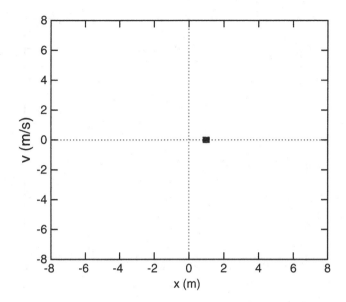

Fig. 5.15 Initial phase space locations for an ensemble of particles in the simple harmonic oscillator system

```
(%i) k:1$ m:1$ dt:0.2$ nt:100$
     array(xre,nt,1000)$       array(vre,nt,1000)$
     for j:0 while j < 1000 do
       block(xre[0,j]:0.8+0.4*random(1.0),
         vre[0,j]: 0.4*random(1.0)-0.2)$
     for i:0 while i < nt do
       block(for j:0 while j<1000 do
         block(xre[i+1,j]: xre[i,j] + dt*vre[i,j],
         vre[i+1,j]: vre[i,j] - k*xre[i,j]*dt/m))$
```

We can plot the resulting collection of points at any of the times we have computed. The code below shows how to construct a list of ordered pairs (x, v) for the initial locations in phase space, as well as a list of the final locations at $t = 20$ s. Figure 5.15 shows a plot of the initial list, and Fig. 5.16 shows the plot of the final list. (The wxdraw2d commands used to generate these plots should be familiar by now, so that code has been suppressed. Likewise, we will leave out the wxdraw2d commands for all code in the remainder of this section.)

```
(%i) xvi:makelist([xre[0,j],vre[0,j]],j,0,999)$
     xvf:makelist([xre[nt,j],vre[nt,j]],j,0,999)$
```

A comparison of Figs. 5.15 and 5.16 shows that the phase space area covered by our ensemble of orbits is growing. However, *Liouville's Theorem* states that the phase space area should remain constant for a conservative system such as the simple harmonic oscillator. Another way to think about how the Euler algorithm will affect the phase space area of an ensemble in phase space is to consider how

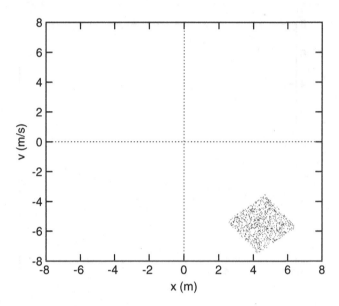

Fig. 5.16 Phase space locations at $t = 20$ s for the ensemble shown in Fig. 5.15, as determined by the Euler algorithm

the infinitesimal volume element, $dx\,dv$, changes under application of the algorithm. We can view the mapping $(x_n, v_n) \rightarrow (x_{n+1}, v_{n+1})$ as a change of variables. The new volume element, $dx_{n+1}dv_{n+1}$, is related to the old volume element, $dx_n dv_n$, by the equation

$$dx_{n+1}dv_{n+1} = |\det J|dx_n dv_n, \qquad (5.24)$$

where J is the Jacobian matrix defined by

$$J = \begin{bmatrix} \dfrac{\partial x_{n+1}}{\partial x_n} & \dfrac{\partial x_{n+1}}{\partial v_n} \\ \dfrac{\partial v_{n+1}}{\partial x_n} & \dfrac{\partial v_{n+1}}{\partial v_n} \end{bmatrix}. \qquad (5.25)$$

If the algorithm preserves the area occupied by an ensemble of orbits in phase space, then the absolute value of the determinant of this Jacobian will be equal to one. If the absolute value of the determinant is not equal to one, then the algorithm fails to preserve phase space area. It is not hard to show that the determinant of the Jacobian for the SHO Euler algorithm is greater than one, thus indicating that the phase space area will grow under repeated applications of the algorithm.

All of these problems indicate that the Euler algorithm is unstable for this system. Although it starts off close to the exact solution, it does not remain there. This is a general property of the algorithm. We can improve its performance by making the time step smaller (requiring that we increase the number of time steps used to cover the same range of time). Reducing the time step will keep the Euler solution close to the exact solution for a longer period of time, but the Euler solution will still diverge after a sufficiently long time. To deal with this problem we need to introduce another algorithm.

To further explore the Euler algorithm for this system, try reducing the time step to 0.05 s and repeating the above computations. Increase the value of nt so that your solution reaches $t = 20$ s. Does reducing the time step help? Does the reduction of the time step solve the problems with the Euler algorithm?

5.4.2   The Euler–Cromer Algorithm

The Euler–Cromer algorithm is a minor modification of the Euler algorithm. The modification consists of updating the velocity first, and then using the new velocity to update the position:

$$v_{n+1} = v_n + F_n \Delta t/m, \tag{5.26}$$

$$x_{n+1} = x_n + v_{n+1} \Delta t. \tag{5.27}$$

This apparently minor change makes a big difference. Below we use the Euler–Cromer algorithm to calculate x and v using the same parameters that we used for the Euler algorithm above. We then construct the same set of plots that were constructed for the Euler algorithm, providing a direct comparison of the performance of the two algorithms (the code to construct the plots is omitted since it is essentially identical to the code used to construct the plots for the Euler algorithm results). Figure 5.17 shows the Euler–Cromer values for $x(t)$ as + marks, with the exact solution appearing as a gray line.

```
(%i) k:1$ m:1$ dt:0.2$ nt:100$
      array(xc,nt)$ array(vc,nt)$
      xc[0]:1$ vc[0]:0$
        for i:0 while i < nt do
        block(vc[i+1]:vc[i]-k*xc[i]*dt/m,
        xc[i+1]:xc[i]+vc[i+1]*dt)$
```

Compared to the results of the Euler algorithm, the improvement is striking. As Fig. 5.18 demonstrates, the velocity comparison is even more impressive.

The phase space plot shown in Fig. 5.19 does not reveal any tendency of the Euler–Cromer values to spiral away from the true values. The Euler–Cromer solution does not coincide precisely with the exact solution, but it remains close to the exact solution and does coincide with that solution at four times during each oscillation.

The Euler–Cromer algorithm does a much better job of conserving energy, as Fig. 5.20 shows. Although the approximated energy value oscillates, the amplitude of these oscillations is very small and the average energy over one period of the oscillator remains fixed (and is close to the correct value, in this case 0.5 J).

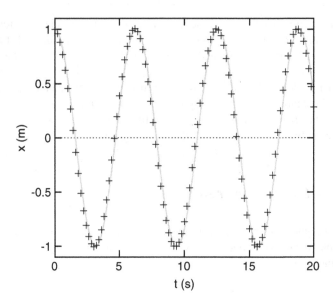

Fig. 5.17 Euler–Cromer approximation (*pluses*) and exact solution (*thick gray curve*) for $x(t)$ for the simple harmonic oscillator

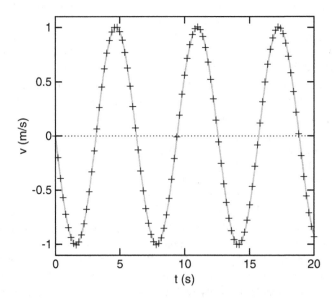

Fig. 5.18 Euler–Cromer approximation (*pluses*) and exact solution (*thick gray curve*) for $v(t)$ for the simple harmonic oscillator

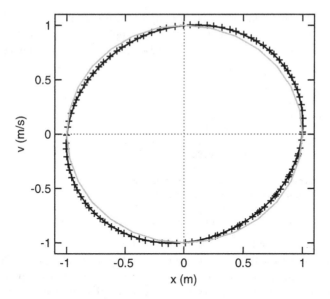

Fig. 5.19 Euler–Cromer approximation (*pluses*) and exact solution (*thick gray curve*) for the phase space trajectory of the simple harmonic oscillator

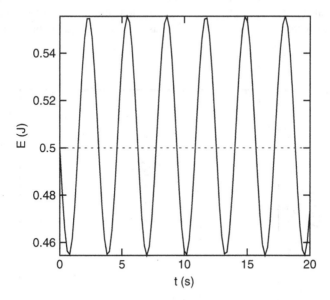

Fig. 5.20 Euler–Cromer approximation for $E(t)$ for the simple harmonic oscillator. The correct energy for this oscillator is 0.5 J

Finally, we examine the question of the preservation of phase space area. First, we initialize our ensemble of particles and then apply the Euler–Cromer algorithm to the ensemble using nested for ...while ...do loops, as shown in the code below.

```
(%i) k:1$ m:1$ dt:0.2$ nt:100$
      array(xrc,nt,1000)$
      array(vrc,nt,1000)$
      for j:0 while j < 1000 do
         block(xrc[0,j]:0.8+0.4*random(1.0),
         vrc[0,j]:  0.4*random(1.0)-0.2)$
      for i:0 while i < nt do
         block(for j:0 while j<1000 do
           block(vrc[i+1,j]: vrc[i,j] -
             k*xrc[i,j]*dt/m,
           xrc[i+1,j]: xrc[i,j] + dt*vrc[i+1,j]) )$
```

Next we plot the resulting ensembles at $t = 0$ and at $t = t_{nt}$. We can create lists of the (x, v) points for the initial ensemble and the ensemble at $t = 20$ s, using essentially the same code that was used for this purpose in our investigation of the Euler algorithm. Plots of these data lists show the region of phase space occupied by our ensemble at each time. Figure 5.21 shows the initial positions, and Fig. 5.22 shows the positions at $t = 20$ s.

We see that the region occupied by the ensemble of orbits has moved and changed its shape, but that it does not appear to have changed its size. This suggests that the Euler–Cromer algorithm preserves phase space area (and thus conforms to Liouville's Theorem).

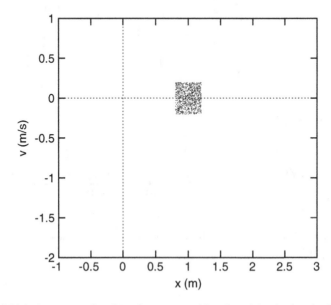

Fig. 5.21 Initial phase space locations for an ensemble of particles in the simple harmonic oscillator system

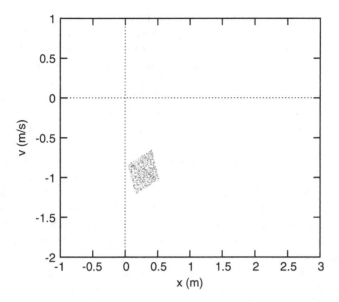

Fig. 5.22 Phase space locations at $t = 20$ s for the ensemble shown in Fig. 5.15, as determined by the Euler–Cromer algorithm

You may want to experiment further with the Euler–Cromer algorithm for this system by reducing the time step. Do the Euler–Cromer algorithm results fit better to the exact solution with a smaller time step? What happens to the energy plot?

5.4.3 Comparing Algorithms

The preceding sections introduced two algorithms for numerically solving systems of ODEs, the Euler algorithm and the Euler–Cromer algorithm. Examining the behavior of these algorithms introduces Liouville's Theorem and the preservation of phase space area (as well as how to determine whether a given algorithm preserves phase space area). In addition to these topics a few other important ideas merit our attention.

To solve a system of ODEs numerically, a computer must use an algorithm. For a given system, some algorithms are better than others. For example, the Euler–Cromer algorithm is clearly superior to the Euler algorithm for oscillatory systems like the SHO. For other systems, however, the Euler–Cromer algorithm will not work very well. In fact, no single algorithm is best for all problems. Choosing the best available algorithm is an important part of numerically solving a system of ODEs.

Which algorithm one chooses may depend on the properties of the system being analyzed. If the phase space dynamics are not of interest, then it may not be very important that the algorithm preserves phase space area. An algorithm that produces a very slowly increasing or decreasing energy (rather than an oscillating energy) but doesn't preserve phase space area might be preferred. The examples above demonstrate that even the Euler–Cromer algorithm doesn't do a perfect job of getting $x(t)$ and $v(t)$ right: the error in those functions grows over time (this is because the Euler–Cromer solution has a slightly different period from the exact solution, so it goes in and out of phase with the exact solution over time). The selection of an algorithm depends on what is important for the calculation you are trying to perform.

Generally, reducing the time step improves the performance of ODE solving algorithms, at a cost: reducing the time step requires more steps (and more computations) to cover the same time interval. For difficult problems the added time for computation can become problematic. Ideally, we want an algorithm that performs well even with a large time step. The most sophisticated algorithms adapt the size of the time step to make it as large as possible while still achieving the desired precision.

Often we cannot know that we have chosen the best algorithm for a particular problem. The only way to proceed is to test the algorithm. Check for certain things like conservation of energy or phase space area to see if the algorithm is unstable. If the algorithm appears to be stable then we can estimate the error produced by the algorithm by comparing the results obtained using one time step with those obtained using a smaller (say, by a factor of 2) time step. The difference between the two provides an estimate for the error in the numerical solution. This procedure will not only indicate whether or not the algorithm is acceptable, it will also help with the selection of the time step. Again, we want to use the largest time step that will produce a solution with the required precision.

For more discussion of ODE solver algorithms, see Sect. A.3.

5.5   Exercises

1. Use a `for` loop to generate the cubes of the first six multiples of five. Do this in two ways:

   - with a `step` command to generate the multiples of five, and
   - without a `step` command.

2. Think carefully about what happens in *Maxima* when you run the following code:

   ```
   x:5$ for i:0 while x<3000 do x:x*x-14$
   display(x)$
   ```

   Fill in the table below to show what happens before, during, and after each pass through the `for` loop.

i	x before loop	is $x < 3000$?	x after loop

What value of x will be displayed when this code is run? (You may want to check your answer.)

3. Write a program in which you specify the integers n and r (with $n \geq r$) and then the program computes the value of $_nC_r = \frac{n!}{r!(n-r)!}$. Do not use any of *Maxima*'s built-in functions (like `factorial` or `binomial`) to do this. Just use simple arithmetic expressions and programming with `for` and `if` ...`then`.

4. Define the following piecewise function in *Maxima*:

$$f(x) = \begin{cases} x^2, & x < 6, \\ 54 - 3x, & 6 \leq x \leq 10, \\ 2x^2 - 40x + 224, & x > 10. \end{cases} \qquad (5.28)$$

Create a plot of this function over the range $0 < x < 15$. Is this function continuous? Is it everywhere differentiable?

5. Modify the code for estimating π in Sect. 5.2.1 to instead estimate the fraction of area within the unit square (with $0 < x < 1$ and $0 < y < 1$) that satisfies the condition $y < 5x - 3$. Plot $y = 5x - 3$ over an appropriate range and comment on whether your result seems reasonable based on the plot.

6. In Sect. 5.2.2 we looked at the motion of an ensemble of particles experiencing constant acceleration (as in freefall near Earth's surface). If these particles were subject to linear air resistance, their distances of fall (x) and velocities (v) are given by:

$$\begin{aligned} x(t) &= x_0 + g\tau t + (v_0 - g\tau)\tau(1 - e^{-t/\tau}), \\ v(t) &= g\tau + (v_0 - g\tau)e^{-t/\tau}, \end{aligned} \qquad (5.29)$$

where $\tau = m/b$ is the characteristic time associated with the air resistance on these particles. Suppose $\tau = 1\,\text{s}$.[6] Modify the *Maxima* code in Sect. 5.2.2 so that it displays the motion of particles falling with air resistance. Use the same range of initial values. Plot the original ensemble (at $t = 0$) and the ensemble at $t = 1$ s. Does the area occupied by this ensemble remain constant? (Note: if you aren't sure, try plotting the ensemble at later times to see if the area changes noticeably over longer intervals.)

---

[6]This is totally unrealistic. A typical value for, say, a raindrop would be more like 10^{-5} s.

7. Examine a new version of the random walk from Sect. 5.2.3 with $p(+x) = 0.26$, $p(-x) = 0.24$, $p(+y) = 0.23$, and $p(-y) = 0.27$. What are the expected values of x and y after N steps for this walk? Modify your code so that you can generate the plots for this walk. Examine walks of different lengths ($N = 200, 2000$, 10,000, and 20,000). Do your results fit with these expected values? What is the expected distance from the origin after N steps? Make a plot of distance as a function of step number and compare the results to the expected distance.

8. Consider a different version of a random walk. The walker begins at the origin. At each time step the walker chooses a direction at random (i.e., the walker chooses an angle $0 < \theta < 2\pi$, with equal probability for any angle in that range). The walker then takes a step of unit length in that direction. Generate plots of this random walk, as well as of the distance from the origin as a function of the number of steps. (Note: you may want only to go up to 2000 steps, since the evaluation of trig functions can bog *Maxima* down.) Compare this random walk to the standard random walk discussed in Sect. 5.2.3. What are the expected values of x and y after N steps of this walk? Make a plot of distance from the origin as a function of step number for this walker. How does this plot compare to the same plot for the standard random walk discussed in the text?

9. Use the Newton–Raphson method to find the two roots of the function $f(x) = \sin(x) - x^2 + \log(x) + 2$ on the interval $(0, 4]$. Determine the values of the roots to six decimal places. You may want to start by plotting this function in order to determine approximate locations of the roots.

10. A damped harmonic oscillator with $\omega_0 = 10$ rad/s and $\beta = 0.5$ s^{-1} is launched from its equilibrium position ($x = 0$) with speed $v_0 = 1$ m/s. Use the results from Sect. 4.4 to construct $x(t)$ for this oscillator. Find $v(t)$, then use the Newton–Raphson algorithm to determine the first time after $t = 0$ at which the speed of the oscillator is half of its initial speed. (Note: you should plot $v(t)$ first.) What is the position of the oscillator at this time? What fraction of the oscillator's initial energy has been lost by this time?

11. Calculate the determinant of the Jacobian for the Euler algorithm applied to the simple harmonic oscillator. Explain how your result relates to the increase in phase space area occupied by the ensemble of particles shown in Figs. 5.15 and 5.16. Then calculate the determinant of the Jacobian for the Euler–Cromer algorithm applied to the simple harmonic oscillator. Explain how your result relates to the preservation of phase space area occupied by the ensemble of particles shown in Figs. 5.21 and 5.22.

12. Repeat the investigation of the Euler–Cromer algorithm for the simple harmonic oscillator in Sect. 5.4, but this time include a damping force $F = -bv$ with $b = 0.1$ kg/s. Redo all of the calculations and plots from that section that related to the Euler–Cromer algorithm, but with this new damping force included. Comment on how the damping force alters the results. Do the results make physical sense? What happens to the phase space area occupied by an ensemble of particles in this system? Calculate the determinant of the Jacobian for the Euler–Cromer algorithm as applied to this damped harmonic oscillator. How does this result relate to your results for the phase space area occupied by an ensemble of particles?

13. Consider the quartic oscillator with potential energy function $V(x) = \alpha x^4$.

   (a) What is the force, as a function of x, on a particle in this potential?
   (b) Use the Euler–Cromer algorithm to solve the equations of motion for the quartic oscillator with $m = 1$ g, $\alpha = 1$ ergs/cm^4, $x(0) = 1$ cm and $v(0) = 0$. Use a time step of 0.01 s and integrate the equations of motion from $t = 0$ to $t = 10$ s. Construct plots of position versus time, velocity versus time, and the trajectory in phase space (velocity versus position). Discuss how the motion of this oscillator compares to that of a harmonic oscillator.
   (c) Use your plots to estimate the period of the oscillations. (Note: if you completed Problem 8 in Sect. 4.8, compare your answer to your result for that problem.)
   (d) Construct a plot of energy versus time for the quartic oscillator, using your Euler–Cromer algorithm results. Does the Euler–Cromer algorithm conserve energy (on average)? Should it conserve energy in this case?
   (e) Construct the Jacobian matrix for the Euler–Cromer algorithm, as applied to the quartic oscillator. Find the determinant of this matrix. Is it equal to one? Will the Euler–Cromer algorithm preserve phase space area? If not, what will happen to the area over time as we apply the algorithm?
   (f) Plot an ensemble of points in the phase space at $t = 0$ and then show where these points end up at $t = 10$ s using the Euler–Cromer algorithm with a time step of 0.1 time units. Do your results show that phase space area is preserved, or does it seem to grow or decrease? Does this agree with your answer to the previous question?

14. Consider a projectile moving in the x–y plane subject to both gravity ($F_g = -mg\hat{y}$) and linear air resistance ($F_r = -b\vec{v}$).

   (a) Write down the x- and y-components of the net force on this projectile. Then write down Newton's Second Law for this projectile as a system of four first-order ODEs.
   (b) Write down the Euler–Cromer algorithm for this system.
   (c) Use your Euler–Cromer algorithm to numerically solve the equations of motion for initial conditions $x(0) = 0$, $y(0) = 0$, $v_x(0) = 10$ m/s, and $v_y(0) = 15$ m/s. Let $m = 1$ kg, $g = 9.8$ m/s^2, and $b = 0.1$ N s/m. Use a time step of 0.01 s. Construct plots of x versus t, y versus t, y versus x, and v_y versus t. Make sure you extend your solution until the projectile hits the ground. Comment on how your results here differ from the results you would expect with no air resistance.
   (d) Construct a plot of v_y versus y. Comment on how the shape of this plot is altered by air resistance.

Chapter 6
Nonlinearity and Chaos

One of the most exciting areas of current research in classical mechanics is the dynamic behavior of nonlinear systems. Nonlinear systems exhibit a much richer variety of behaviors than do linear systems. However, most nonlinear systems are impossible to solve using paper and pencil methods. It was only with the advent of digital computers that nonlinear dynamics really came into its own. With computer software like *Maxima* we can explore nonlinear dynamics in a way that is impossible without a computer.

6.1 Nonlinear Dynamics

To understand what *nonlinear dynamics* means, we first define what we mean by *linear*. To this point, we have examined systems whose dynamics are defined by differential equations of the form

$$\ddot{x} = f(x, \dot{x}, t). \tag{6.1}$$

Such a system is considered *linear* if the dependent variable x and its derivatives appear only to the first power (i.e., contain only linear terms). In other words, the equation of motion does not contain any factors of x^2, or $1/x$, or $\cos(\dot{x})$, etc. If the equation of motion in Eq. 6.1 is to be linear, it must have the form

$$\ddot{x} = a(t)\dot{x} + b(t)x + c(t) \tag{6.2}$$

where $a(t)$, $b(t)$, and $c(t)$ are functions that depend only on the time t. The harmonic oscillator systems that we studied in the previous chapter are all linear systems in this sense.

© Todd Keene Timberlake & J. Wilson Mixon, Jr. 2016
T.K. Timberlake, J.W. Mixon, *Classical Mechanics with Maxima*, Undergraduate
Lecture Notes in Physics, DOI 10.1007/978-1-4939-3207-8_6

A *nonlinear* system is just a system whose equation of motion contains one or more nonlinear terms. A simple example that we have already encountered is a particle in free fall with quadratic air resistance, which has the equation of motion

$$\ddot{x} = -g - c\dot{x}^2. \tag{6.3}$$

The $\dot{x}^2$ term makes this equation, and thus the system, nonlinear.

Why is the distinction between linear and nonlinear important? It is because linear systems obey something called the *superposition principle*, while nonlinear systems do not. The superposition principle states that if $x_1(t)$ is a solution to a particular linear differential equation, and $x_2(t)$ is a different solution to the same equation, then $ax_1(t) + bx_2(t)$ is also a solution of that equation for any constants a and b. But this result doesn't hold for nonlinear systems. We can find two independent solutions to a nonlinear differential equation, but a linear combination of these solutions may not be a solution to that equation.

As mentioned above, nonlinear systems can exhibit a rich variety of behaviors that are not seen in linear systems. One particularly remarkable type of motion that can occur in nonlinear, but not in linear, systems is *chaotic* motion. Chaotic motion is characterized by a sensitive dependence on initial conditions, in which a slight change in the initial conditions can lead to dramatically different motion after some time. Chaotic systems cannot be solved (even approximately) using analytical methods, so very little was known about chaotic systems until the development of computers that could generate approximate numerical solutions for these systems.

This chapter provides an introduction to how *Maxima* can be used to explore nonlinear dynamics and chaos in a few example systems. These examples illustrate many important features of nonlinear dynamics and chaos, but this chapter should not be taken as a comprehensive treatment of the subject. For more on this exciting area of classical mechanics, see the references at the end of this chapter.

6.2   The van der Pol Oscillator

6.2.1   The Undriven Case

We begin our investigation of nonlinear systems by examining a modified harmonic oscillator system. This system is a version of the "van der Pol oscillator." It consists of an oscillator with a linear restoring force (like the harmonic oscillator) as well as a nonlinear force. The dimensionless equations of motion for the system are

$$\dot{x} = v, \tag{6.4}$$

and

$$\dot{v} = -x + \gamma(1 + x^2 - v^2)v, \tag{6.5}$$

where x is a dimensionless position variable, v is a dimensionless velocity, and γ is a constant.

Fig. 6.1 Phase space trajectory for a Van der Pol oscillator starting from rest at $x = 0.5$. The x and v coordinates are dimensionless

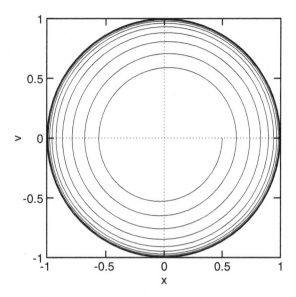

We solve for the motion of this system numerically using $\gamma = 0.1$ and two different initial conditions. We begin by finding the motion with initial conditions $x_0 = 0.5$ and $v_0 = 0$. The code below uses *Maxima*'s rk command to generate 1000 values of t, x, and v. The code then generates a data list containing the ordered pairs of x and v values for various times. The last portion of the code generates a plot of the phase space trajectory (v versus x) for our van der Pol oscillator, which appears as Fig. 6.1.

```
(%i) gamma:0.1$                 data:
     rk([v,-x+gamma*(1-x^2-v^2)*v], [x,v],[0.5, 0],
         [t,0,100,0.1] )$
       xvL: makelist([data[i][2], data[i][3]],i,1,
           length(data))$
       wxdraw2d( user_preamble = "set size ratio 1",
       xlabel="x", ylabel ="v", xaxis = true,
          yaxis = true, xtics = 1/2, ytics = 1/2,
          point_size=0,points_joined=true,points(xvL))$
```

We see that this trajectory spirals outward, gradually approaching a circle with unit radius. Next we examine the motion when the initial conditions are $x_0 = -1.5$ and $v_0 = 0$. The phase space trajectory is shown in Fig. 6.2. The code to generate the solution and plot is almost identical to the code above, except for the change in the initial value for x and the use of the line_type=dots option in wxdraw2d in order to plot the trajectory as a dotted line. Figure 6.2 shows that for this initial condition the trajectory spirals inward, again approaching a circle with a unit radius.

Our two trajectories both approach the unit circle $x^2 + v^2 = 1$. To illustrate this we plot this circle as a thick line (line_width=2) and then combine it with the two previous plots. The resulting plot is shown in Fig. 6.3.

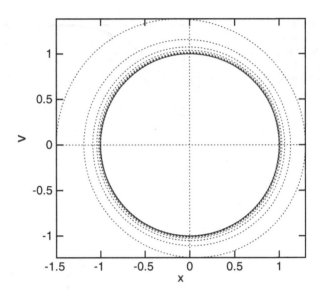

Fig. 6.2 Phase space trajectory for a Van der Pol oscillator starting from rest at $x = -1.5$. The x and v coordinates are dimensionless

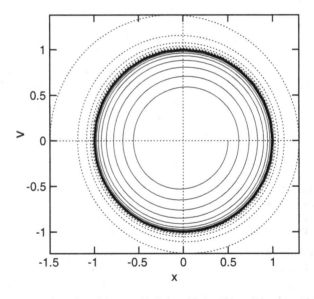

Fig. 6.3 Phase space trajectories of the van der Pol oscillator (*thin solid curve* and *dotted curve*) shown with the unit circle (*thick solid curve*)

This convergence to the unit circle illustrates the self-limiting nature of the Van der Pol oscillator. The initial motion depends upon the initial conditions, but eventually the motion settles into a particular pattern (clockwise rotation on the unit circle) regardless of the initial conditions. The pattern of motion into which the system settles is known as a *limit cycle* or *attractor*. We have already seen an example of a limit cycle in the case of the driven harmonic oscillator, but in the van der Pol oscillator this limit cycle behavior occurs without a driving force.

The *Maxima* command `drawdf` is useful for analyzing dynamical systems like the van der Pol oscillator. This command produces a plot of the "direction field" for a system of ODEs. The direction field shows the direction of motion for a particle at a grid of points in the state space. The `drawdf` command can also be used to display a particular trajectory (with specified initial conditions) along with the direction field. The basic arguments of the `drawdf` (or `wxdrawdf` for producing plots inside a *wxMaxima* notebook) consist of three lists. The first list contains the equations that specify the derivatives for each of the two system variables. The second list specifies one of the system variables and the maximum and minimum values of that variable to be used in the plot. The third list is the same as the second, but for the other system variable. The `drawdf` command allows a variety of options as well. The *Maxima* manual provides more information on `drawdf`.

The commands below generate the direction field for the van der Pol oscillator with $\gamma = 0.3$. The results appear in Fig. 6.4. Note that to use the `drawdf` command we must first load the `drawdf` package. The `soln_at` option generates a solution curve that passes through the specified point. Here we generate two different solution curves. One solution passes through the point ($x = 0.5, v = 0$) and the other passes through ($x = 1.5, v = 0$). The `duration` option controls the length of the time interval over which the solutions are generated. Note that the solution curves cover a time interval that starts before, and ends after, the time when the curve passes through the specified location.

Fig. 6.4 Direction field plot for the van der Pol oscillator with $\gamma = 0.3$. Two trajectories are shown: the *thick curve* passes through ($x = 0.5, v = 0$) and the *thin curve* passes through ($x = 1.5, v = 0$). The x and v coordinates are dimensionless

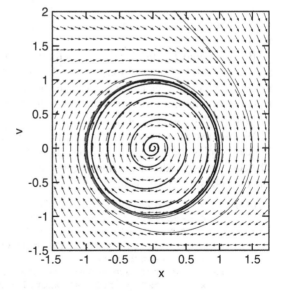

```
(%i) load(drawdf)$      gamma:0.3$
     wxdrawdf([v,-x+gamma*(1-x^2-v^2)*v],[x,-1.5,1.75],
     [v,-1.5,2.0], color=black,line_width=2,
     duration=20, proportional_axes=xy,soln_at(0.5,0),
     line_width=1, soln_at(1.5,0),xlabel="x",
     ylabel="v", dimensions=[480,480])$
```

Note how the trajectories follow the direction of the arrows in the direction field. The behavior of the van der Pol oscillator for $\gamma = 0.3$ is similar to that for $\gamma = 0.1$, examined above. The two trajectories spiral outward or inward, as needed, to reach the unit circle. The only difference is that with $\gamma = 0.3$ the trajectories approach the unit circle faster than they did for $\gamma = 0.1$. *Exercise*: Use drawdf to examine the behavior for other values of γ.

6.2.2 The Driven Case

The van der Pol oscillator exhibits limit cycle behavior even without a driving force. What happens if we *do* add a driving force to this oscillator? The dimensionless equations of motion for this system are

$$\dot{x} = y, \tag{6.6}$$

and

$$\dot{v} = -x + \gamma(1 + x^2 - v^2)v + \alpha\cos(\omega t), \tag{6.7}$$

where α represents the amplitude and ω represents the frequency of the driving force.

Below we find the numerical solution for this system with $\alpha = 0.9$, $\gamma = 0.25$ and $\omega = 2.47$ with initial conditions $x_0 = -1.5$ and $v_0 = 0$. We then use the solution to generate a plot of the phase space trajectory in Fig. 6.5. The code to produce the solution and plot requires only a small modification of the code given above for the undriven system, so it is not shown here.

We can also view the phase space trajectory in three dimensions, as in Fig. 6.6. We can display the x–v plane horizontally and allow the time axis to run upward. The code below produced the numerical solution, generates the necessary data list, and then plots this 3D trajectory.

```
(%i) w:2.47$   data3:rk([v,-x+0.25*(1-x^2-v^2)*v+
        0.9*cos(w*t)],[x,v],[-1.5,0],
        [t,0,200*%pi/w,0.1])$
     xvt3d:makelist([data3[i][2],data3[i][3],data3[i][1]],
        i, 1, length(data3)))$
     wxdraw3d(user_preamble="set size ratio 1",
        xlabel="x",ylabel="v",zlabel="t",xaxis=true,
        yaxis=true, xtics = 1/2, ytics = 1/2,
```

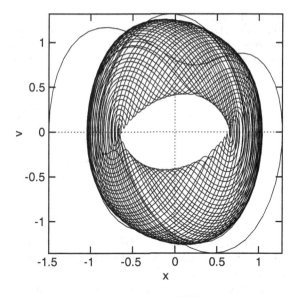

Fig. 6.5 Phase space trajectory for a driven van der Pol oscillator. The x and v coordinates are dimensionless

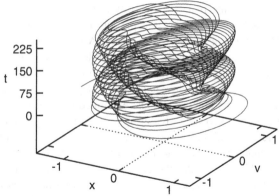

Fig. 6.6 Three-dimensional plot of the location in phase space as a function of time for the driven van der Pol oscillator. The x, v, and t coordinates are dimensionless

```
ztics=75, point_size = 0,
points_joined = true, points(xvt3d),
dimensions=[480,480])$
```

The 3D plot in Fig. 6.6 shows some initial transient behavior, after which the motion seems to settle into a pattern. It is hard to tell, though, whether the pattern is repeating (periodic motion) or not (quasi-periodic motion).

We gain a different perspective on the motion by looking at the location of the particle in the x–v plane each time the driving force completes one cycle. In other words, we find the intersection between the 3D trajectory shown above (with the time axis running upward) and a plane of constant time (a horizontal plane, in this case). Figure 6.7 illustrates this concept by displaying the 3D trajectory along with

Fig. 6.7 The phase space
path shown in Fig. 6.6
superimposed with a plane of
constant time at $t = 100\pi/\omega$

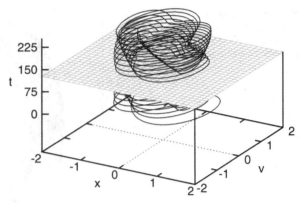

the plane representing $t = 100\pi/\omega$.[1] The black representation of the trajectory
cuts the gray representation of the plane from below at approximately $x = -0.69$,
$y = 0.72$).

Now we look at *only* those points where the trajectory intersects one of the
constant time planes defined by $t = 2\pi n/\omega$, where n is an integer. That way we
are only seeing where the particle is in the x–v plane at the end (or beginning)
of each cycle of the driving force. A plot of these intersection points is known as
a *strobe plot* (because we are only looking at the trajectory at discrete, regularly
spaced times—much like watching a moving object that is illuminated by a strobe
light). The code below generates a strobe plot for the driven van der Pol oscillator
using the parameters we just examined (the wxdraw command has been omitted).
Figure 6.8 shows the resulting plot. Note that the time step for the rk command is
one tenth of the period of the driving force. This time step is sufficiently small
so that we can get accurate results from the Runge–Kutta algorithm. (You can
test this by redoing the calculation with a smaller time step to see if it makes a
difference.) However, we only want to plot points at increments of the full period,
so the makelist command for constructing the xvL4 list uses only every tenth
element of the data4 list, so only the points with $t = 2\pi n/\omega$, where n is an integer,
appear in the plot.

```
(%i)  w:2.47$ data4:rk([v,-x + .25*(1-x^2-v^2)*v+
       0.9*cos(w*t)],[x,v], [-1.5, 0],
       [t,0,200*%pi/w,0.2*%pi/w])$
     xvL4:makelist( [data4[i][2], data4[i][3]],
       i,1,length(data4),10)$
```

We see that all but a few points lie on a closed curve in the phase space. We can
show that these few points that don't lie on the closed curve are associated with the

---

[1] To get a better view of the intersection of the trajectory and the plane, remove the wx from this
command and execute it. A graph appears outside the *wxMaxima* session. Expand and rotate that
graph.

Fig. 6.8 Strobe plot for the
driven van der Pol oscillator.
The x and v coordinates are
dimensionless

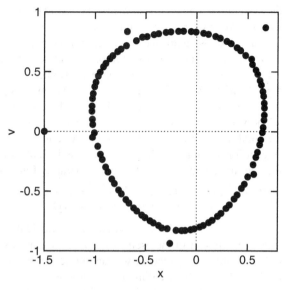

Fig. 6.9 Same as Fig. 6.8 but
with the first ten points
omitted to eliminate transient
motion

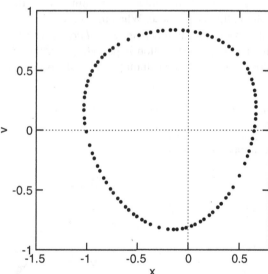

transient motion during the first few cycles. We can do this by plotting only those
points from the tenth cycle onward. Figure 6.9 shows the resulting plot.

Once we get past the transient behavior at the beginning, the strobe plot reveals a
simple pattern. All points lie on a closed curve in the phase space. This closed curve
serves as the *attractor* for the strobe plot. However, there are an infinite number
of points on this attractor. The system *does not* repeat itself, so the motion is not
periodic. However, the motion is restricted such that at the beginning of each cycle

the system must be on this attractor. This kind of motion is known as *quasiperiodic motion*. We saw another example of quasiperiodic motion when we examined the 2D harmonic oscillator with incommensurate frequencies in Sect. 4.3.

We can construct a strobe plot for the undriven van der Pol oscillator that appears earlier in this section. We do not have a periodic driving force, so we must determine the period at which we sample the data. An intuitive choice would be to sample at the natural frequency of the oscillator without the nonlinear term. For the version of the van der Pol oscillator that we have been studying, that frequency would be 2π (in dimensionless units). Now we can construct a strobe plot at this frequency. Except for the function that generates the data, all commands are as above, so only the command that includes the data-generating function appears below.

```
(%i) data5:rk([v,-x+0.25*(1-x^2-v^2)*v],[x,v],[-1.5,0],
        [t,0,200*%pi,0.628318531])$
```

The plot in Fig. 6.10 looks odd until we realize that what we are seeing is just the transient behavior of the oscillator as it approaches its limit cycle. If we want to see only the limit cycle behavior we can ignore the first 90 oscillations and then plot the data for the next 10, yielding the plot in Fig. 6.11.

As the oscillator settles into its limit cycle, the strobe plot settles into a single point in the phase space. When the attractor for a strobe plot is a single point in the state space, we call this point a *fixed point*. A fixed point in the strobe plot indicates that the long-term motion is periodic, with the period used to generate the strobe plot (in this case, the natural period of the oscillator without the nonlinear term).

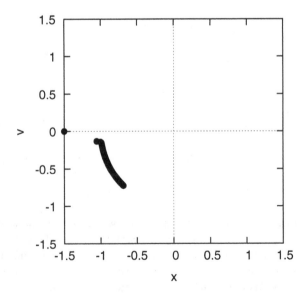

Fig. 6.10 Strobe plot for the undriven van der Pol oscillator. The x and v coordinates are dimensionless

Fig. 6.11 Same as Fig. 6.10 but with the first 90 points omitted to eliminate transient motion

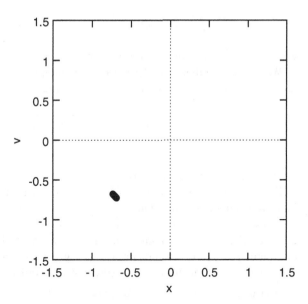

6.3   The Driven Damped Pendulum

The van der Pol oscillator is not the first nonlinear oscillator we have encountered. In Sect. 4.7 we examined the simple pendulum. The simple pendulum is an example of a nonlinear oscillator because the restoring force is proportional to the sine of the position angle, not to the position angle itself. In the limit of small oscillations the pendulum closely approximates a linear (harmonic) oscillator. But we have already seen that for large oscillations the simple pendulum can deviate from the behavior expected of a linear oscillator.

6.3.1   Solving the Driven Damped Pendulum

The plane pendulum becomes even more interesting when we add damping and driving forces. The equation of motion for a simple pendulum of length L and mass m with linear damping and driven by a sinusoidal force with amplitude F_0 and frequency ω is

$$mL^2\ddot{\phi} = -bL^2\dot{\phi} - mgL\sin\phi + LF_0\cos(\omega t), \tag{6.8}$$

where b is the coefficient for the linear damping.

We can recast this equation of motion in terms of the constants $\beta = b/(2m)$ and $\gamma = F_0/(mg)$. The equation of motion can be written as

$$\ddot{\phi} + 2\beta\dot{\phi} + \omega_0^2 \sin\phi = \gamma\omega_0^2 \cos(\omega t), \qquad (6.9)$$

where $\omega_0 = \sqrt{g/L}$ is the natural frequency for small oscillations of the pendulum.

We must solve this second order differential equation numerically. To do so we rewrite it in the form of two first order differential equations:

$$\dot{x} = v, \qquad (6.10)$$

$$\dot{v} = -2\beta v - \omega_0^2 \sin x + \gamma\omega_0^2 \cos(\omega t), \qquad (6.11)$$

where we use the notation $x = \phi$.

We then solve this system of ODEs numerically using *Maxima*'s rk command. First we examine the case with $\omega = 2\pi$ rad/s (so the period is 1 s), $\omega_0 = 1.5\omega$, $\beta = \omega_0/4$, and $\gamma = 0.2$, with initial conditions $x(0) = 0$ and $y(0) = 0$. The code below generates the numerical solution, organizes the resulting data into a list of ordered pairs (t, x), and creates a plot, Fig. 6.12, of angular position versus time over ten cycles of the driving force (the first 1000 ordered pairs).

```
(%i) w:2*%pi$ w0:1.5*w$ b:w0/4$ g:0.2$
     data:rk([v,-2*b*v-w0*w0*sin(x)+g*w0*cos(w*t)],
        [x,v], [0,0], [t,0,200,0.01])$
     txL:makelist([data[i][1],data[i][2]], i,1,1000)$
     wxdraw2d( xaxis = true, yaxis = true, xtics=1,
        ytics=0.1, xlabel="t (s)",ylabel="x (rad)",
        point_size=0,
        points_joined=true,points(txL))$
```

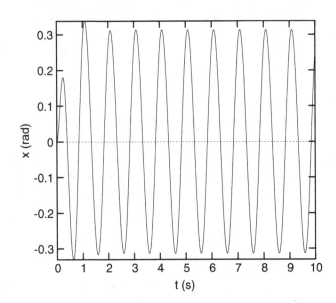

Fig. 6.12   Angular position as a function of time for a driven damped pendulum with $\gamma = 0.2$

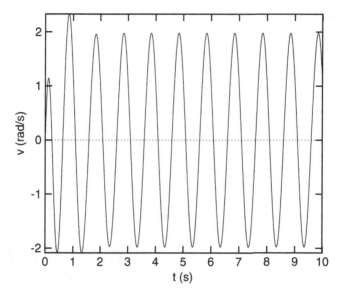

Fig. 6.13 Angular velocity as a function of time for a driven damped pendulum with $\gamma = 0.2$

Fig. 6.14 Phase space
trajectory (angular velocity
versus angular position) for a
driven damped pendulum
with $\gamma = 0.2$

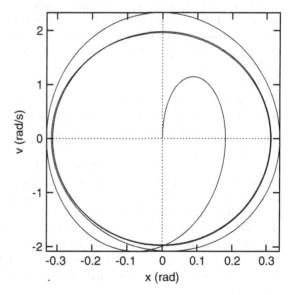

Using similar code we generate a list of ordered pairs (t, v) and plot the angular
velocity versus time over the first ten cycles of the driving force. The resulting plot
is shown in Fig. 6.13.

Finally, Fig. 6.14 shows a plot of the trajectory in phase space (angular velocity
versus angular position) using the full set of data points generated by the rk
command.

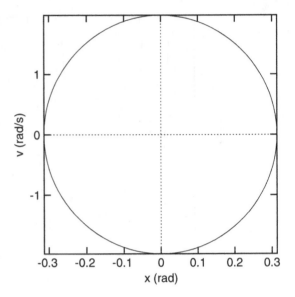

Fig. 6.15 Same as Fig. 6.14 but with the first 5000 data points (up to $t = 5$ s) omitted to eliminate transient motion

The graphs show that after some initial transient behavior the pendulum settles into a periodic oscillation that looks like simple harmonic motion. In other words, the driven damped pendulum at these parameter values exhibits limit cycle behavior just like the driven damped harmonic oscillator and the van der Pol oscillator. We can focus on the limit cycle motion by ignoring the first 5000 data points. The results appear in Fig. 6.15.

This path in the phase space looks just like simple harmonic motion. We check on the periodicity of the pendulum by constructing a strobe plot of the motion, plotting x versus v after each cycle of the driving force. Since the driving force has a period of 1 s, and the time step we used in rk was 0.01 s, we extract every 100th data point to construct our strobe plot, which appears in Fig. 6.16.

We see multiple points, but some of these may reflect transient behavior that occurs before the limit cycle is reached. For example, the point at the origin is just our initial condition. We can exclude the transient behavior by plotting only the points generated after 50 cycles of the driving force have been completed (i.e., ignoring the first 5000 data points from rk). Figure 6.17 shows the result. Now the strobe plot produces a single *fixed point* near $(x = 0.27, v = 1.02)$, indicating that the long-term behavior of the driven damped pendulum with these parameter values is periodic.

6.3.2 Period Doubling

Is this fixed point behavior the only kind of motion that we can get from the driven damped pendulum? We can show that it is not by looking at the motion with different parameter values. We use the same parameter values as above, but with a stronger

Fig. 6.16 Strobe plot of the driven damped pendulum with $\gamma = 0.2$. The coordinates are angular position and angular velocity

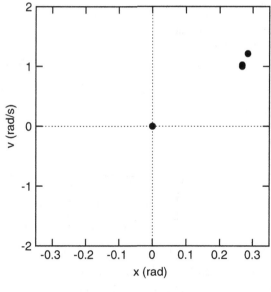

Fig. 6.17 Same as Fig. 6.16 but with the first 5000 data points (up to $t = 5$ s) omitted to eliminate transient motion

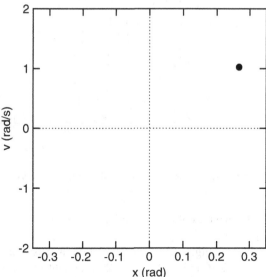

driving force ($\gamma = 1.077$). Also, we start the system at $x = -\pi/2$ and $v = 0$. As before, we begin by examining the angular position over the first ten periods. A plot of the position as a function of time appears in Fig. 6.18.[2]

---

[2]The code for producing this plot and the other plots in this section is just a minor modification of the code shown above, so we do not include the code here.

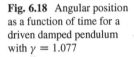

Fig. 6.18 Angular position as a function of time for a driven damped pendulum with $\gamma = 1.077$

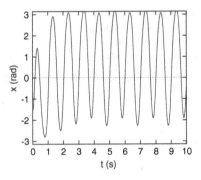

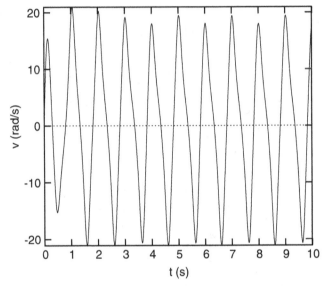

Fig. 6.19 Angular velocity as a function of time for a driven damped pendulum with $\gamma = 1.077$

We next examine the angular velocity over these ten periods, as shown in Fig. 6.19.

Finally, Fig. 6.20 shows the motion in phase space.

Figures 6.18, 6.19, and 6.20 highlight two important aspects of this system's motion. First, the oscillations are no longer sinusoidal. This is clear from the plots of v versus t and v versus x. Second, if we look closely we can see that the motion is not quite periodic, either, at least not with the period of the driving force. The peaks and troughs of the x versus t curve do not always occur at the same values, but seem to hop back and forth between two different values. We examine this behavior in more detail by plotting the phase space trajectory without the transient behavior. Figure 6.21 replicates Fig. 6.20 but starting at $t = 50$ s instead of $t = 0$.

The limiting motion of the pendulum now consists of two different oscillations, repeated one after the other. We can best illustrate this by constructing a strobe

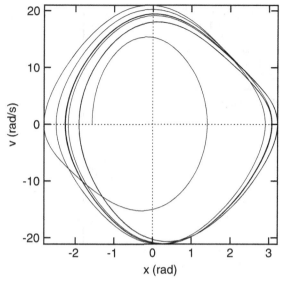

Fig. 6.20 Phase space trajectory (angular velocity versus angular position) for a driven damped pendulum with $\gamma = 1.077$

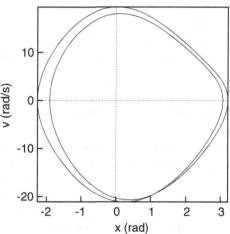

Fig. 6.21 Same as Fig. 6.20 but starting at $t = 50$ s to eliminate transient motion

plot that samples the phase space location after each cycle of the driving field (but ignoring the initial transient motion). The strobe plot in Fig. 6.22 confirms that the pendulum settles into a periodic oscillation, but that the period of this oscillation is not the period of the driving force. Rather, it is twice the period of the driving force. The motion of the pendulum with these parameter values is an example of a *period-2 attractor*. (The fixed point motion we saw earlier can also be referred to as a *period-1 attractor*.) The transition from period-1 to period-2 motion as we

Fig. 6.22 Strobe plot for the driven damped pendulum with $\gamma = 1.077$. The coordinates are angular position x and angular velocity v. The first 50 cycles have been omitted to eliminate transient motion

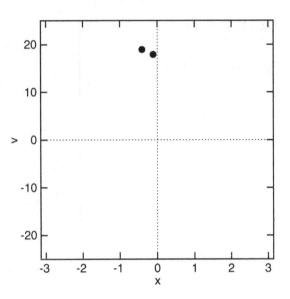

change the value of γ is known as *period doubling*. You can show that the driven damped pendulum undergoes more period doublings, as the value of γ is increased, by examining the motion for $\gamma = 1.081$ and $\gamma = 1.0826$.

6.3.3 Rolling Motion

Another type of behavior that the driven damped pendulum can exhibit is *rolling motion*. Rolling motion occurs when the pendulum swings all the way around. In this case the values of x will go outside the range $[-\pi, \pi]$. We illustrate rolling motion by looking at our driven damped pendulum system with $\gamma = 1.4$ and initial conditions $x(0) = -\pi/2$ and $v(0) = 0$ (all other values as before). Figure 6.23 shows a plot of the angular position as a function of time.

This plot of x versus t shows that in addition to oscillating, the values of x decrease steadily over time. The pendulum is repeatedly swinging all the way around in the clockwise direction. We next examine plots of the angular velocity as a function of time and the phase space trajectory for this pendulum, shown in Fig. 6.24.

The angular velocity values oscillate, while the angular position values oscillate and also decrease over time. Again, this shows that the pendulum is swinging all the way around in the clockwise direction.

Because x is an angle variable, x and $x + 2n\pi$ represent the same angular position if n is an integer. We can account for this equality in our plot by plotting the *principal value* of x, instead of x itself. The principal value is the angle equivalent to x that lies in the principal domain $[-\pi, \pi]$. We can compute the principal value of x by using *modular arithmetic*.

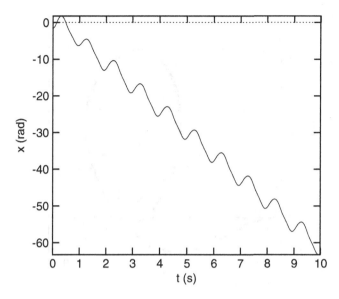

Fig. 6.23 Angular position as a function of time for a driven damped pendulum with $\gamma = 1.4$

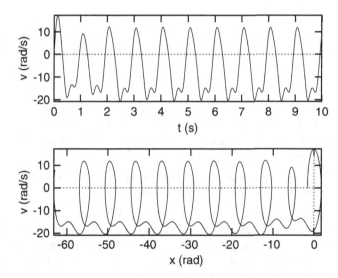

Fig. 6.24 Plots of angular velocity as a function of time (*top*) and phase space trajectory (*bottom*) for a driven damped pendulum with $\gamma = 1.4$

The expression "*a* modulo *b*" means the remainder obtained when *a* is divided by *b*. For example, 12.3 modulo 5 would be 2.3 (because 12.3 divided by 5 is 2 with a remainder of 2.3). If we compute *x* modulo 2π we will get an angle equivalent to *x* in the range $[0, 2\pi]$. This is close to what we want, but not quite it. We can't

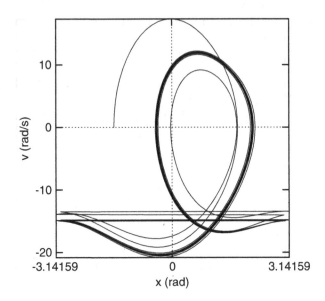

Fig. 6.25 Same as the bottom of Fig. 6.24, but using the principal value of x

just subtract π from our result because doing so will give us an angle that is not equivalent to x. So here's what we do: if x_p represents the principal value of x then

$$x_p = ((x + \pi) \bmod 2\pi) - \pi. \tag{6.12}$$

In *Maxima* we write the expression on the right-hand side as mod(x + %pi, 2*%pi) - %pi. The code below generates a list of ordered pairs (x_p, v). (Note that data3 is the list containing the output from rk for this system.) We can then construct a plot of v versus x_p, as shown in Fig. 6.25 (code omitted).

```
(%i) xvLr3:makelist([mod(data3[i][2]+%pi,2*%pi)-%pi,
        data3[i][3] ],i,1,length(data3))$
```

Figure 6.25 illustrates the rolling motion of the pendulum. The pendulum eventually settles into a path that enters the plot from the right side, does a loop in the phase space, and then goes out of the plot on the left side. Because the left side ($x = -\pi$) is an angle equivalent to the right side ($x = \pi$), when the trajectory goes out the left side it immediately comes in on the right side.

There is one problem with the plot above, though. When the trajectory goes out the left side and comes in the right side, *Maxima* connects the last point on the left side and the first point on the right side with a line. This leads to the horizontal lines in the range $-15 < y < -13$. These lines are artifacts of the way *Maxima* constructs the plot, they are not really part of the trajectory. We can remove these spurious line segments by using the xrange option to truncate the x values a bit. For example, using the option xrange=[-.95*%pi, .95*%pi] results in the plot shown in

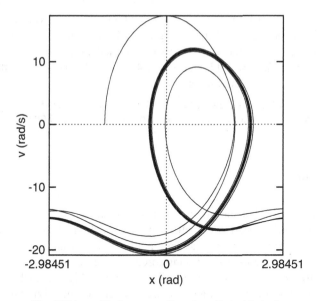

Fig. 6.26 Same as Fig. 6.25 but with the range of x values limited in order to prevent spurious lines

Fig. 6.27 Strobe plot for the driven damped pendulum with $\gamma = 1.4$. The first 50 cycles have been omitted to eliminate transient motion

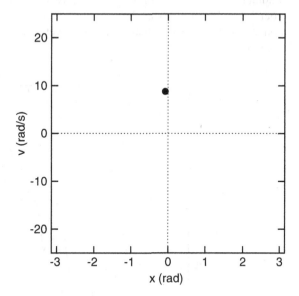

Fig. 6.26. This plot misses a few points on either end but eliminates the spurious lines in Fig. 6.25.

Now we can construct a strobe plot of the motion, using the principal value of x and ignoring the transient motion at the beginning. The code below shows how to construct the list of data points for such a plot. Figure 6.27 shows the plot of this data set. This plot shows that the motion is periodic with the same period as the driving force. So again we have a fixed point, but this time with rolling motion instead of oscillatory motion. This rolling motion goes through a period-doubling sequence just like the oscillatory motion we examined earlier. You can examine the motion for $\gamma = 1.45$, $\gamma = 1.47$, and $\gamma = 1.477$ (all other parameters the same as above) to see this period-doubling in action.

```
(%i) xvLsp3:makelist([mod(data3[i][2]+%pi,2*%pi)-%pi,
            data3[i][3] ], i, 5001, length(data3), 100)$
```

6.3.4 Chaos

So far we have seen that the driven damped pendulum can exhibit motion with a period-1, period-2, period-4, etc. attractor. It can exhibit these periodic motions while oscillating or while rolling (swinging all the way around). What else can it do? To find out, we increase the driving force a bit more, to $\gamma = 1.5$, with initial conditions $x(0) = -\pi/2$ and $v(0) = 0$. We begin with Fig. 6.28, which shows the system's angular position, this time over twenty cycles of the driving force.

The x versus t plot in Fig. 6.28 shows that with these parameter values the pendulum exhibits a kind of behavior that we have not seen before. At first it swings around (rolling motion) in the clockwise direction, but then it begins to swing around counterclockwise before reversing yet again and swinging in a clockwise direction.

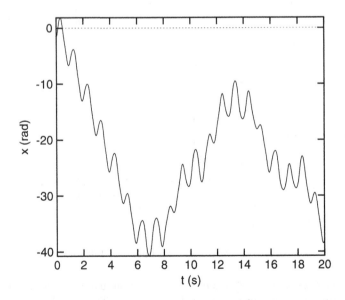

Fig. 6.28 Angular position as a function of time for a driven damped pendulum with $\gamma = 1.5$

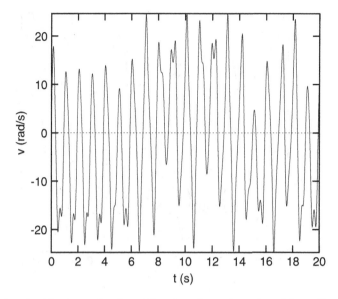

Fig. 6.29   Angular velocity as a function of time for a driven damped pendulum with $\gamma = 1.5$

We next observe the angular velocity v, shown in Fig. 6.29. The angular velocity oscillates, but it never seems to fall into a repeating pattern.

Finally, we observe the trajectory in phase space, as shown in Fig. 6.30. The phase space representation shows the oscillations in v and the rolling motion (with changing directions) in x. This state space trajectory shows no signs of settling in to a limit cycle.

We can reconstruct Fig. 6.30 using the principal value of x rather than x itself. As above, we truncate the xrange in order to prevent spurious horizontal lines. The resulting plot appears in Fig. 6.31.

The motion shown in this plot looks much more complicated than what we have seen before. Could this be quasiperiodic motion, as we saw in the van der Pol oscillator? To find out we construct a strobe plot. To ensure that we have enough data points in our strobe plot we recompute our numerical solution, this time using larger time steps ($\delta t = 0.1$) and examining 5000 cycles of the driving force. We discard the first 100 cycles to eliminate any transient motion from our plot. The resulting plot is shown in Fig. 6.32.

The strobe plot in Fig. 6.32 shows that the motion is definitely not periodic. Neither does it quite look like the quasiperiodic motion we saw in the van der Pol oscillator. The points lie on a curve in the state space, but close inspection reveals that the curve has a complicated structure. It seems to fold in on itself, creating loops within loops. Although we cannot show this numerically, the number of loops within the curve is infinite.

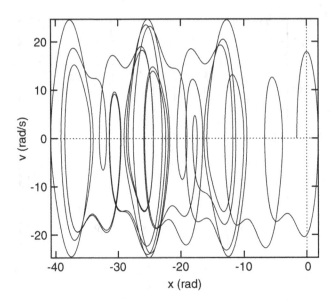

Fig. 6.30 Phase space trajectory for a driven damped pendulum with $\gamma = 1.5$

Fig. 6.31 Same as Fig. 6.30 but using the principal value of x and truncating the x range to avoid spurious lines

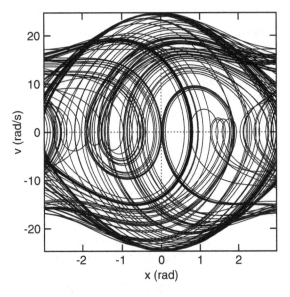

This structure, which has infinite length, differs dramatically from the finite-length quasiperiodic attractor that we saw in the van der Pol oscillator. The structure shown in Fig. 6.32 is known as a *strange attractor*. The points in the strobe plot are restricted to the region of state space occupied by the attractor, but the motion is neither periodic nor quasiperiodic. In fact, we will show that the motion is *chaotic*. Strange attractors are characteristic of chaotic systems with damping.

Fig. 6.32 Strobe plot for the driven damped pendulum with $\gamma = 1.5$. The coordinates are angular position and angular velocity. The first 100 cycles have been omitted to eliminate transient motion

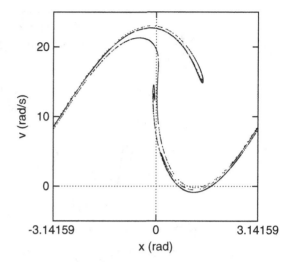

Chaotic motion is motion that exhibits sensitive dependence on initial conditions. This means that if we start the system off with two sets of initial conditions that are almost—but not exactly—identical, then after some time the motions that follow from these two sets of initial conditions are quite different. This is essentially the opposite of limit cycle behavior, because with a limit cycle we find that even widely different initial conditions can still converge to the same limit cycle motion in the long term.

To illustrate sensitive dependence on initial conditions, we use our driven damped pendulum system with $\gamma = 1.5$. This time we look at the motion generated by two different sets of initial conditions. One set of initial conditions will be $x(0) = -\pi/2$, $v(0) = 0$. The other set will be $x(0) = -\pi/2 + 0.001$ and $v(0) = 0$. The two versions of the driven damped pendulum are starting off at almost the same angle (just 0.001 radians difference) and both at rest. Everything else about the two pendulums is the same. The code below generates the numerical solutions for these two pendulums a single plot showing $x(t)$ for both pendulums (some options for the wxdraw2d command have been omitted). The resulting plot is shown in Fig. 6.33.

```
(%i) w:2*%pi$ w0:1.5*w$ b:w0/4$ g:1.5$ x0:-%pi/2$
     data6a: rk( [v,-2*b*v-w0*w0*sin(x) +
        g*w0*w0*cos(w*t)], [x,v],[x0,0],
        [t,0,200*%pi/w,0.01])$
     data6b: rk([v,-2*b*v-w0*w0*sin(x) +
        g*w0*w0*cos(w*t)], [x,v], [x0+0.001,0],
        [t, 0, 200*%pi/w, 0.01])$
     txL6a:makelist([data6a[i][1],
        data6a[i][2] ],i,1,2000)$
     txL6b:makelist([data6b[i][1],
        data6b[i][2] ],i,1,2000)$
     wxdraw2d( ..., key="x(0)=%phi/2,v(0)=0", ...,
        points(txL6a),key="x(0)=%phi/2+.001, v(0)=0",
        ..., points(txL6b) )$
```

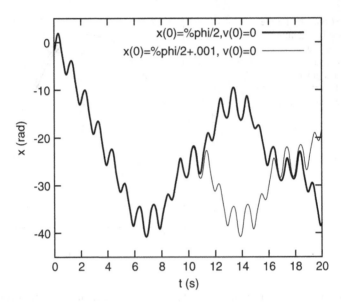

Fig. 6.33 Angular positions as a function of time for driven damped pendulums with $\gamma = 1.5$ and two slightly different initial conditions

The two pendulums exhibit nearly identical motion until about $t = 11$ s. After that time, however, their motions are quite different. This is what is meant by sensitive dependence on initial conditions: the two versions of the system start with nearly identical initial conditions but end up behaving quite differently in the long term.

What happens if we start our two pendulums off even closer together, say with a difference in x of only 0.0001 radians? Modify the code above and generate the new plot. You should find that starting the pendulums closer together only delays the inevitable. Their motion stays matched until about $t = 14$ s, but then deviates noticeably. Starting the pendulums off only 0.00001 radians apart keeps the motion the same until about $t = 17$ s. In fact, each factor of ten reduction in the difference in $x(0)$ results in the motion staying the same for only a few additional cycles of the driving force.

To repeat, this sensitive dependence on initial conditions is characteristic of chaotic systems. This behavior has important consequences for the predictability of these systems. It is nearly impossible to predict the long-term behavior of a chaotic system. Our knowledge of the initial conditions of any system is inevitably limited in precision. In non-chaotic systems this lack of precise knowledge may not matter, since a slight difference in initial conditions will produce only a slight difference in motion. But in chaotic systems a slight difference in the initial conditions will eventually produce a large difference in the motion.

A more precise determination of the initial conditions doesn't help. As we just saw in the driven pendulum, reducing the uncertainty in our initial conditions by a factor of ten (a big improvement!) will only allow us to make accurate predictions of the motion for a slightly longer time. The fact that *dividing* the difference in $x(0)$ by a certain factor only *adds* a little bit to the time the trajectories stay close to each other suggests that there is some sort of exponential relationship between the difference in x and the elapsed time. In fact, another way to define "sensitive dependence on initial conditions" is to say that nearby trajectories diverge exponentially in time. Mathematically we can express this behavior as

$$|\Delta x(t)| \approx |\Delta x(0)|e^{\lambda t}, \tag{6.13}$$

where $\Delta x(t)$ represents the difference in x for our two pendulums and λ is called the "Lyapunov exponent."

The Lyapunov exponent characterizes how rapidly the exponential divergence takes place (large λ indicates a very rapid exponential divergence, while small λ indicates a slower exponential divergence). The trajectories *will* diverge exponentially as long as λ is positive. Systems with non-chaotic dynamics will have $\lambda = 0$, indicating that the trajectories either maintain a fixed separation or else diverge/converge slower than exponentially (linearly, quadratically, etc.), or $\lambda < 0$, indicating that the trajectories converge exponentially.

We can illustrate the exponential divergence of trajectories in the driven damped pendulum by constructing a plot of $\log(|\Delta x|)$ versus t. If the trajectories diverge exponentially we should find

$$\log(|\Delta x(t)|) \approx \lambda t + \log(|\Delta x(0)|). \tag{6.14}$$

Our plot should show that $\log(|\Delta x(t)|)$ increases linearly with time, at least when averaged over sufficiently long times. The slope of the line is the Lyapunov exponent. The code below illustrates how to construct this plot for the two pendulums shown in Fig. 6.33. The resulting plot is shown in Fig. 6.34.

```
(%i) logD:makelist([data6a[i][1],log(abs(data6a[i][2]-
     data6b[i][2]))],i,1, 2000)$
     wxdraw2d( ..., points(logD) )$
```

Figure 6.34 indicates that, on average, $\log(|\Delta x(t)|)$ does increase linearly over time. Nearby trajectories (such as this pair, with an initial divergence of only 0.001 radians) diverge exponentially, the Lyapunov exponent is positive, and the motion is chaotic.

The value of $\log(|\Delta x(t)|)$ oscillates. We should expect this because both pendulums are oscillating. There will be times when they happen to pass by each other, even if their motion is very different. But this plot shows that over long times the pendulums do in fact get farther apart, exponentially fast.

What would a plot of $\log(|\Delta x|)$ versus t look like for the other (non-chaotic) parameter values that we studied? You can find out for yourself by modifying the code shown above.

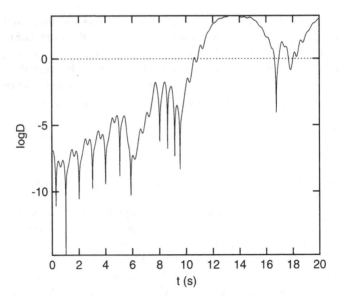

Fig. 6.34 Plot of $\log D = \log(|\Delta x|)$ as a function of time (in s) for the pendulums shown in Fig. 6.33. The curve roughly follows an upward sloping line. The slope of this line is the Lyapunov exponent

6.4   Maps and Chaos

We have seen that the driven damped pendulum has some interesting characteristics: oscillatory and rolling motion, with a period-1 attractor (or fixed point in the strobe plot). Furthermore, as the strength of the driving force is increased, it exhibits a succession of period-doublings that eventually culminates in chaotic motion on a strange attractor. Exploring these characteristics in greater detail (both numerically and analytically) is instructive, but the numerical solution of the ODEs is so time-consuming that a more detailed exploration may not be practical. Fortunately, a more easily solved system that exhibits these same characteristics is available.

6.4.1   The Logistic Map

We wish to examine chaotic dynamics in a system with *deterministic dynamics*, such that the state of the system at time $t + \Delta t$ is determined by the state of the system at time t. In most physical systems the deterministic dynamics is generated by a set of differential equations. It is possible, however, to generate deterministic dynamics in a simpler way, using an *iterated map*. An iterated map generates a sequence of numbers (or ordered pairs of numbers, etc.). Each term in the sequence is a function of the preceding term. We have already seen iterated maps used in the context of the Newton–Raphson method for finding roots in Sect. 5.3.

As an example of an iterated map, consider the "logistic equation" defined by

$$f(r,x) = rx(1-x). \tag{6.15}$$

This function can be used to define an iterated map that generates a sequence of values x_n such that

$$x_{n+1} = f(x_n) = rx_n(1-x_n). \tag{6.16}$$

This *logistic map* is used in studies of predator–prey population dynamics, but we are primarily interested in this iterated map because it generates behaviors that are very similar to those exhibited by the driven damped pendulum. These behaviors can be studied more easily in the context of an iterated map than they can in the context of differential equations.

To get an idea of how the logistic map works, consider what happens when we choose a particular value for r and a particular initial value for x. Suppose we choose $r = 0.$ and $x_0 = 0.2$. The successive values of the sequence will be 0.2, 0.08, 0.0368, 0.0177, etc. The code below uses the `for` loop introduced in Sect. 5.1 to generate the list of x values produced by iterating the logistic map for any value of r and any initial value of x. We can then plot the resulting series of values to examine the behavior of the map. Figure 6.35 shows the result of using `draw` to create this plot.

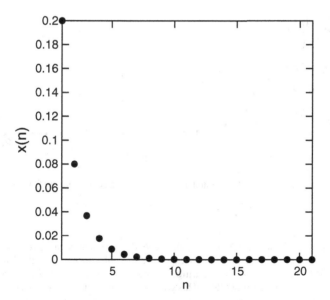

Fig. 6.35 Sequence of values generated by the logistic map with $r = 0.5$

```
(%i) f(x, r) := r*x*(1-x)$
     x0:0.2$ data: [x0]$
        for i thru 20 do block(
        x0:f(x0,0.5),
        data: append(data, [x0]))$
     wxdraw2d(xlabel="n", ylabel="x(n)",
        point_type=7,point_size=1,points(data))$
```

Maxima also has built-in functions that can display the dynamics of an iterated map in various ways. These functions are part of the dynamics package. To use these functions we must first load this package. The evolution command (part of the dynamics package) generates a plot of the sequence of values produced by an iterated map. The syntax is evolution($f(x)$, x_0,n) where $f(x)$ is the mapping function, x_0 is the initial value for x, and n is the number of iterations to be performed.

The code below generates a plot of the sequence generated by our logistic map for $r = 0.5$ with $x_0 = 0.2$. The plot, shown in Fig. 6.36, displays in a separate window, rather than within the *wxMaxima* notebook.[3]

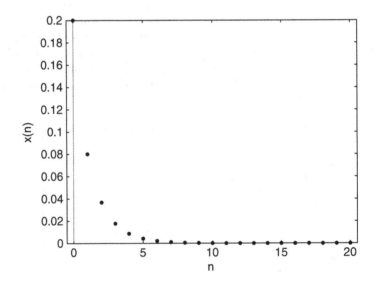

Fig. 6.36 Same as Fig. 6.35, but generated using the evolution command from the dynamics package

[3]Macintosh users need to have *X11* or another xWindows system running in order to display these plots. One can instruct *Maxima* to export the graphic image to a file. The command evolution(f(x,0.5,0.2,20, [gnuplot_term, png]) would place a file named maxplot.png in the user's root folder. This graphic file can be viewed, edited, and incorporated into documents. The *Maxima* manual's plot documentation provides more options that allow controlling the plot and the folder to which it is exported. We copied the plot to a paint program and reduced it to grayscale for inclusion in this book.

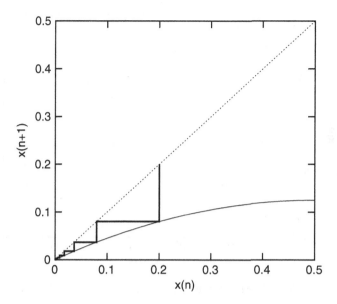

Fig. 6.37 Staircase plot for the logistic map with $r = 0.5$. The *thin, solid curve* is the mapping function. The *dotted line* is $x_{n+1} = x_n$. The *thick line* is the staircase path showing the evolution of the system

```
(%i) load("dynamics")$
     evolution(f(x,0.5),0.2,20,[ylabel, "x(n)"])$
```

We can get a better picture of this iterative process by plotting the logistic equation and the reference line $y = x$. Refer to Fig. 6.37. Begin at x_0 on the line $y = x$. Move vertically to the logistic curve. The y-coordinate of this point is the next term in the sequence. This y-coordinate will now become the x-coordinate used to determine the next term. To accomplish this on our plot we just move horizontally to the line $y = x$. Then we move vertically to the logistic curve again. Again, the y-coordinate of this point is the next term of the sequence. If we continue this procedure indefinitely we get the picture of the entire sequence shown in Fig. 6.37. The code below shows how to produce this plot.

```
(%i) f(x, r) := r*x*(1-x)$
     x0:0.2$ data: [[x0,x0]]$
     for i thru 200 do block(
         x1:f(x0,0.5),
         data:append(data,[[x0,x1],[x1,x1]]),x0:x1)$
     wxdraw2d(xlabel="x(n)", ylabel="x(n+1)",
        explicit(f(x,0.5),x,0,0.5),
        line_type=dots,
        explicit(x,x,0,.5), points_joined=true,
        line_width=3, line_type=solid, point_type=0,
        points(data))$
```

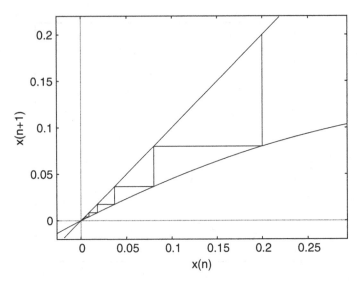

Fig. 6.38 Same as Fig. 6.37, but generated with the `staircase` command from the `dynamics` package

Note how the thick line, which traces the evolution of the sequence, bounces between the dotted line (the $y = x$ line) and the thin, solid line (the mapping function). The plot clearly illustrates that the sequence produced by the logistic map, with $r = 0.5$ and $x_0 = 0.2$, converges steadily to zero. This kind of plot is sometimes called a "staircase plot" or a "cobweb plot." Fortunately, the `dynamics` package provides a way of generating these plots without having to program a `for` loop. The command for generating these plots is `staircase` and the syntax is almost identical to that for `evolution`. The code shown below produces the staircase (or cobweb) plot for our logistic map with $r = 0.5$ and $x_0 = 0.2$ that appears in Fig. 6.38.

```
(%i) staircase(f(x,0.5),0.2,20,
          [xlabel,"x(n)"],[ylabel,"x(n+1)"])$
```

As noted before, this staircase plot shows that the sequence of numbers converges to zero. For the logistic map with $r = 0.5$, $x = 0$ is a *fixed point* (or period-1 attractor, sometimes called a "1-cycle") of the map. What if we change the value of r? The staircase plot for $r = 2$ (with $x_0 = 0.2$) is shown in Fig. 6.39. (The command, essentially identical to the preceding one, is omitted.)

When $r = 2$, the map again converges, but to 0.5 instead of to zero. We will examine a few more values of r to see how changing r affects the system's behavior. Figure 6.40 shows the map when $r = 2.9$.

When $r = 2.9$ the sequence still converges, to $x = 0.655$. Now the convergence occurs much more slowly than it did for $r = 2$ or $r = 0.5$. Instead of approaching the fixed point directly, the staircase curve gradually spirals in toward the fixed point.

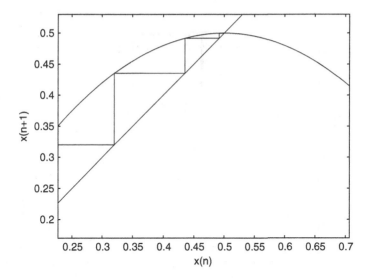

Fig. 6.39 Staircase plot for the logistic map with $r = 2$

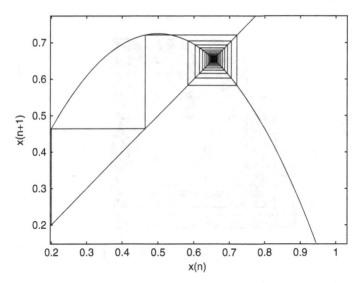

Fig. 6.40 Staircase plot for the logistic map with $r = 2.9$

Figure 6.41 shows the staircase plot for $r = 3.1$. When r increases to 3.1, the sequence does not converge onto a single value. Rather, the sequence eventually oscillates between two values (0.558 and 0.765). This is an example of a 2-cycle (or period-2 attractor). This change in behavior (from converging to a 1-cycle to converging to a 2-cycle) is an example of a period-doubling, just like those we saw in the driven damped pendulum.

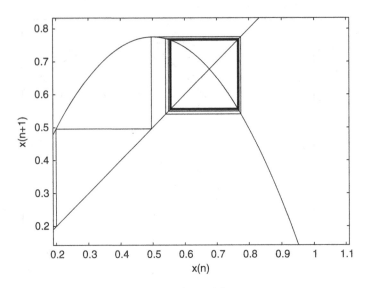

Fig. 6.41 Staircase plot for the logistic map with $r = 3.1$

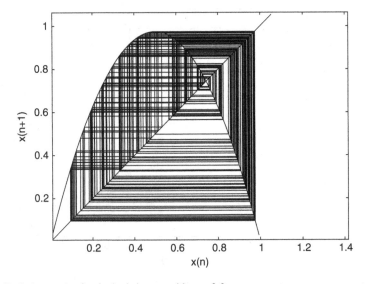

Fig. 6.42 Staircase plot for the logistic map with $r = 3.9$

Figure 6.42 shows the staircase plot for $r = 3.9$. For $r = 3.9$ the sequence does not seem to converge to any value, or to any repeating series of values. The system now exhibits *chaotic dynamics*. The sequence of x values for this map seems to be a list of essentially random numbers. This appearance is deceptive, however. This system is still deterministic, not random.

Examining the logistic map shows that it reproduces many of the behavioral features that we observed in the driven damped pendulum: periodic attractors, period-doubling, and chaos. More importantly for our purposes, the relative simplicity of iterated maps lets us delve more deeply into the dynamics of the logistic map.

6.4.2 Bifurcation Diagrams

To get a more detailed picture of what happens in this system as r is increased, we can create a **bifurcation diagram**. For several different values of r, and for some initial value x_0, we generate the sequence of x's using our map. We can then plot all of these values of x versus r, but we leave out the first several hundred (or so) x's for each r value. That way we are only seeing the behavior of the map after it has converged (if it does converge), not the transient behavior at the beginning of the sequence.

Bifurcation diagrams can be constructed using the orbits command from the dynamics package. The syntax is

$$\text{orbits}(f(x, r), x_0, n_1, n_2, [r, r_{\min}, r_{\max}], \text{options}).$$

As before, f is the mapping function and x_0 is the initial value of x used to generate each sequence. Here n_1 is the number of x values at the beginning of the sequence for each r value that are discarded, while n_2 is the number of subsequent x values to be plotted. The r values to be used range from $r_{\min}$ to $r_{\max}$.

For the logistic map defined by Eq. 6.15 we do not consider values of r greater than 4 or less than 0 because such values of r can lead to negative values in our sequence and the map can go off to infinity. The code below creates a bifurcation diagram for the logistic map. The option nticks specifies the number of r values to be used in generating the plot. Note that the code may take a while to run. After all, it has to generate $250 \times 400 = 100{,}000$ x values! Figure 6.43 shows the resulting bifurcation diagram.

```
(%i) orbits(f(x,r), 0.1, 50, 200, [r, 0, 4],
            [style, dots], [nticks,400])$
```

In the diagram above we see that for $r < 1$ the map has a 1-cycle at $x = 0$. For $2 < r < 3$ the map still has a 1-cycle but the value of x for the 1-cycle increases from 0 up to 2/3. At $r = 3$ the 1-cycle disappears and for $3 < r < 3.45$ the map has a 2-cycle. This change from a 1-cycle to a 2-cycle at a particular parameter value is known as a period-doubling *bifurcation*. (This particular bifurcation is an example of a *pitchfork bifurcation*.) Another period-doubling bifurcation occurs at $r = 3.45$ leading to a 4-cycle. The sequence of period-doubling bifurcations continues until about $r = 3.57$, at which point the map becomes chaotic. Even for $r > 3.57$ there are some particular values of r that lead to regular N-cycles, rather than chaotic behavior (most notably near $r = 3.84$).

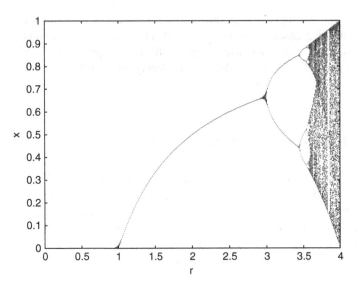

Fig. 6.43 Bifurcation diagram for the logistic map with $0 \leq r \leq 4$

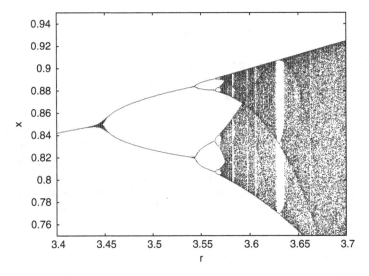

Fig. 6.44 Detail of the bifurcation diagram for the logistic map with $3.4 \leq r \leq 3.7$

We can use the `orbits` command to zoom in on a particular region of the bifurcation diagram. For example, we could focus on the top branch of the diagram shown above (with $0.75 < x < 0.95$) in the parameter range $3.4 \leq r \leq 3.7$ using the code below. Figure 6.44 shows the result.

```
(%i) orbits(f(x,r), 0.224, 100, 400, [r, 3.4, 3.7],
        [style, dots],[nticks,400],[x,0.75,0.95])$
```

A comparison between the detail plot in Fig. 6.44 and our original bifurcation diagram in Fig. 6.43 reveals an interesting new feature: the bifurcation diagram is *self-similar*. An object is self-similar if a small part of the object looks like the entire object. Here we see that our detailed look at a small portion of the bifurcation diagram looks very much like the entire bifurcation diagram. Self-similarity is a characteristic of *fractals*, a fascinating topic that is unfortunately beyond the scope of this book.

6.4.3 Diverging Trajectories

As we did with the driven pendulum, we can investigate how nearby trajectories in the logistic map system diverge. First consider a map that converges to a 1-cycle ($r = 2.9$) for initial conditions $x_0 = 0.1$ and 0.3. The code below generates a single plot showing both of these two sequences. We will refer to the curves as *trajectories* with different initial conditions. Figure 6.45 shows the plot of these trajectories.

```
(%i) f(x, r) := r*x*(1-x) $
     x0:0.1$ data1: [x0] $
     for i thru 50 do block(
        x0:f(x0,2.9),
        data1: append(data1,[x0]))$
     f(x, r) := r*x*(1-x) $
     x0:0.3 $ data2: [x0] $
     for i thru 50 do block(
```

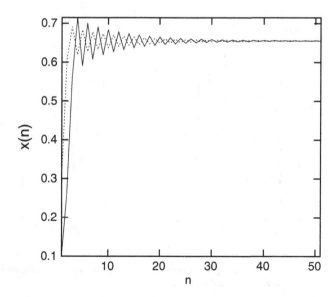

Fig. 6.45 Two trajectories for the logistic map with $r = 2.9$

```
x0:f(x0,2.9),
data2: append(data2,[x0]))$
wxdraw2d(xlabel="n", ylabel="x(n)",
  points_joined=true, point_type=0,points(data1),
  line_type=dots, points(data2))$
```

In Fig. 6.45 the trajectory with $x(0) = 0.1$ is displayed as a solid line, while the trajectory with $x(0) = 0.3$ is displayed as a dotted line. Although the trajectories do not start off close to each other, and they follow different paths for many iterations, eventually they both converge on the same value. This behavior is characteristic of a regular (non-chaotic) map with a periodic attractor. Different trajectories converge into the same trajectory in the long run. They may end up in a 2-cycle, or a 4-cycle, or some other N-cycle, but they all end up doing the same thing.

Next we examine the behavior of trajectories that begin very close together, but with a value of r in the chaotic regime. First we examine the sequences for $r = 3.9$ with $x_0 = 0.2$ (shown as a solid line) and 0.20001 (shown as a dotted line) in Fig. 6.46.

Here we see that the trajectories start off very close together and remain together for about 18 iterations. After 18 iterations, though, the trajectories diverge and become easily distinguishable. As with the driven damped pendulum, this sensitive dependence on initial conditions is characteristic of chaotic systems. We can reduce the difference in initial conditions by a factor of 100 and see what happens. The resulting plot is displayed in Fig. 6.47.

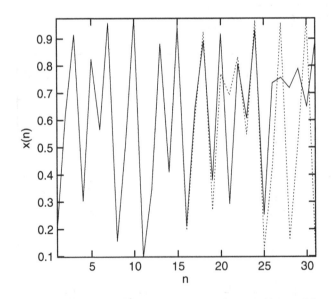

Fig. 6.46 Two trajectories for the logistic map with $r = 3.9$. The initial conditions for the two trajectories are nearly identical (0.2 and 0.20001)

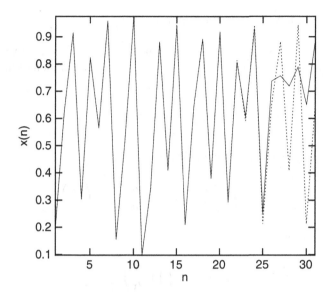

Fig. 6.47 Same as Fig. 6.46 but with initial conditions closer together (0.2 and 0.2000001)

The trajectories *do* stay together longer, but for only about 25 iterations. This is not a big improvement considering how much closer our initial values were. Nearby trajectories diverge at an exponential rate in chaotic systems. We now turn to a closer examination of this exponential divergence.

6.4.4 Lyapunov Exponents

In the logistic map, as in the driven damped pendulum, nearby trajectories will diverge exponentially when the system is chaotic. This means that, on average, the distance between trajectories with initial conditions x_0 and $x_0 + d_0$ after n iterations of the map is

$$d_n = d_0 e^{n\lambda}, \tag{6.17}$$

where λ is the Lyapunov exponent of the map. If λ is negative then the trajectories will converge, as we saw for the logistic map with $r = 0.5$, 2, 2.9, and 3.1. If $\lambda = 0$, then nearby trajectories may converge or diverge but at a rate that is slower than exponential. However, if λ is positive we get the exponential divergence that is characteristic of chaotic maps (and chaotic physical systems).

For the driven damped pendulum we estimated the Lyapunov exponent by graphing the logarithm of the difference in angle between the two trajectories as a function of time. The approximate slope of the curve gave the Lyapunov exponent. For the logistic map we can derive an analytical expression for the Lyapunov exponent.

6.4.4.1   Derivation of Lyapunov Exponent

Suppose we have two trajectories, one of which starts with the initial value x_0 and the other starts with initial value $x_0' = x_0 + \epsilon$. After one iteration of the map we have, to first order in ϵ,

$$x_1' = f(x_0 + \epsilon) \approx f(x_0) + \epsilon \left(\frac{df}{dx}\right)_{x=x_0} = x_1 + \epsilon \left(\frac{df}{dx}\right)_{x=x_0}, \tag{6.18}$$

and after two iterations of the map we have

$$x_2' \approx f\left(x_1 + \epsilon \left(\frac{df}{dx}\right)_{x=x_0}\right) \approx x_2 + \epsilon \left(\frac{df}{dx}\right)_{x=x_1} \left(\frac{df}{dx}\right)_{x=x_0}. \tag{6.19}$$

After n iterations we have

$$x_n' \approx x_n + \epsilon \prod_{i=0}^{n-1} \left(\frac{df}{dx}\right)_{x=x_i}, \tag{6.20}$$

where x_i is the ith iterate of x_0. So the distance (absolute value of the difference) between x_n and x_n' is

$$d_n \approx \left|\epsilon \prod_{i=0}^{n-1} \left(\frac{df}{dx}\right)_{x=x_i}\right| = d_0 \left|\prod_{i=0}^{n-1} \left(\frac{df}{dx}\right)_{x=x_i}\right|, \tag{6.21}$$

where $d_0 = |\epsilon|$.

Comparing this to Eq. 6.17 we find

$$e^{n\lambda} = \left|\prod_{i=0}^{n-1} \left(\frac{df}{dx}\right)_{x=x_i}\right|. \tag{6.22}$$

We can take the natural logarithm of both sides and solve for λ to find

$$\lambda = \frac{1}{n} \sum_{i=0}^{n-1} \ln \left|\left(\frac{df}{dx}\right)_{x=x_i}\right|. \tag{6.23}$$

There is a problem with this expression for λ: it depends on the value of n, whereas we want the Lyapunov exponent to depend only on the map and possibly the initial conditions. Therefore we define the Lyapunov exponent to be the limit of this expression as n goes to infinity:

$$\lambda = \lim_{n\to\infty} \frac{1}{n} \sum_{i=0}^{n-1} \ln \left|\left(\frac{df}{dx}\right)_{x=x_i}\right|. \tag{6.24}$$

If $f(x)$ is a linear function (so that its derivative is constant) then this expression reduces to $\lambda = \ln(|df/dx|)$. Similarly, if x_0 happens to be a fixed point of the map then $\lambda = \ln(|(df/dx)_{x=x_0}|)$.

6.4.4.2 Lyapunov Exponent for the Logistic Map

For a given map we can construct a numerical approximation for the Lyapunov exponent. We can only compute an approximate value because we cannot actually perform the infinite sum in Eq. 6.24. Instead of going to the limit $n \to \infty$ we must be content with using a large, but finite, value for n. Often this is sufficient to give us a good approximation for λ.

To carry out these calculations we first evaluate the derivative df/dx. We can use *Maxima* to assign the derivative of f to a new function df.

```
(%i) diff(f(x,r),x)$        df(x,r) := ''%;
(%o) (df(x,r) := r(1-x) - rx
```

Next we define a function that computes an approximation to the Lyapunov exponent for a given r using Eq. 6.24. We define the function using a for loop to carry out the sum. We sum over the first 200 terms, in hopes that this will be sufficient to accurately approximate the Lyapunov exponent. Once this function is defined we can use it to compute the Lyapunov exponent for any value of r and any x_0.[4] The code below defines this function and demonstrates its use.

```
(%i) lyapunov(r,x0):= block(le:0,
         for i:0 thru 199 do block(
           le:le+log(abs(df(x0,r))),
           x0:f(x0,r)
           ),
         return(le/200) )$
      lyapunov(3.8,0.2);
(%o) 0.40059
```

We see that the Lyapunov exponent is positive for $r = 3.8$, indicating that the dynamics are chaotic for that value of r. We can now use our function for computing the Lyapunov exponent to generate a plot of the Lyapunov exponent as a function of r (using $x_0 = 0.2$ as our initial value for all values of r). *Maxima* has difficulty plotting this function as an explicit function, so instead we generate a list of ordered pairs and use points to generate the plot, as shown in the code below. Figure 6.48 shows the resulting plot.

```
(%i) lelist:makelist([0.004*i,lyapunov(0.004*i,0.2)],
         i,1,1000)$
      wxdraw2d(xlabel="r",ylabel="Lyapunov exponent",
         xaxis=true,points_joined=true, point_type=0,
         points(lelist), yrange=[-3,1])$
```

---

[4]Recall that we must restrict ourselves to $0 < r < 4$ because for values of r outside this range the sequence can run off to infinity, which will cause serious numerical problems.

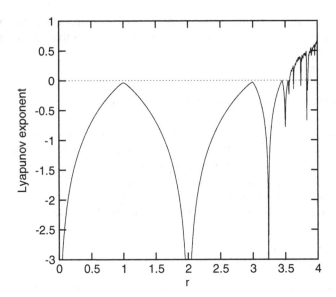

Fig. 6.48 Lyapunov exponent as a function of r for the logistic map

Fig. 6.49 Detail of Fig. 6.48

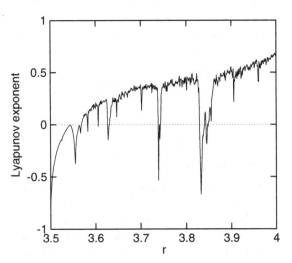

The Lyapunov exponent goes to zero at a bifurcation point (such as at $r = 1$, 3, etc.), but after the bifurcation has occurred it drops below zero again. Also, the Lyapunov exponent becomes positive (indicating a transition to chaos) for some $r > 3.5$. We examine this transition to chaos in more detail by focusing on the region $3.5 \leq r \leq 4$. Figure 6.49 shows a plot of the Lyapunov exponent for this range of r values.

Figure 6.49 shows that the Lyapunov exponent crosses the axis around $r = 3.57$. However, it does occasionally dip below the axis again, such as near $r = 3.63$, 3.74,

3.83, and 3.84. These dips in the Lyapunov exponent correspond exactly to the gaps in the bifurcations diagram for this map. They indicate values of r at which the map is no longer chaotic. Instead, the sequences generated by the map with these r values will converge to a periodic attractor.

6.5 Fixed Points, Stability, and Chaos

The bifurcation diagram for the logistic map shows that the map has a period-1 (or fixed point) attractor that disappears at $r = 3$, giving way to a period-2 attractor. What happens to the fixed point attractor at this point? The answer, in this case, is that the fixed point continues to exist for $r > 3$, but it no longer attracts nearby trajectories. We characterize an attracting fixed point as *stable* and a repelling fixed point as *unstable*. How, other than by examining the bifurcation diagram, can we evaluate the stability of a given fixed point? We addressed this topic briefly in Sect. 5.3, where we showed that the Newton–Raphson method for finding roots of a function employs an iterative map that has a stable fixed point at the root. Here we will take a closer look at the stability of fixed points and illustrate the role that fixed points (and other n-cycles) play in the dynamics of the logistic map.

6.5.1 Stability of Fixed Points

The fixed points of an iterative map are the values of x for which $f(x) = x$, where f is the mapping function. A fixed point $x = x_*$ is stable if the derivative of the mapping function at x_* has absolute value less than one, while the fixed point is unstable if the absolute value of the derivative at the fixed point is greater than one. In summary,

- x_* is a fixed point if and only if $f(x_*) = x_*$,
- x_* is a stable fixed point if $|df/dx|_{x=x_*} < 1$, and
- x_* is an unstable fixed point if $|df/dx|_{x=x_*} > 1$.

The case $|df/dx|_{x=x_*} = 1$ is ambiguous and requires further analysis. Usually fixed points with $|df/dx|_{x=x_*} = 1$ are unstable, but this is a complicated issue that we will not address here.

It is not hard to visualize how these criteria work. A fixed point is simply the intersection of the curve $y = f(x)$ and the line $y = x$. In the vicinity of the fixed point $x = x_*$ we can approximate the function $f(x)$ by the linear function $f_{lin}(x) = x_* + (df/dx)_{x=x_*}(x - x_*)$. This approximate function should accurately describe the behavior of the map for points close to the fixed point. We can explore the stability of a fixed point by generating the staircase plot using f_{lin} and an initial value x_0 that is close to x_*. If the staircase plot displays motion away from the fixed point, then that fixed point is unstable. If the staircase plot converges to the fixed point then that fixed point is stable.

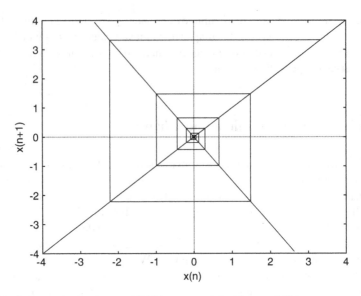

Fig. 6.50 Staircase plot for the case $(df/dx)_{x=x_*} = -1.5$ with $x_0 = 0.001$

We examine staircase plots for four different cases, all with fixed points located at $x_* = 0$ (for convenience). We first consider a case with $(df/dx)_{x=x_*} < -1$ using the code below, which produces the staircase plot in Fig. 6.50 if we use the initial value $x_0 = 0.001$. [5]

```
(%i) load(dynamics)$
     staircase(-1.5*x,0.001, 20, [x,-4,4],[y,-4,4])$
```

According to our criteria above, a fixed point with $(df/dx)_{x=x_*} = -1.5$ should be unstable. Although the sequence shown in Fig. 6.50 starts very close to the fixed point at $x = 0$ it gradually moves away under the action of the mapping. In this case the motion spirals outward in the staircase plot, indicating that the sequence alternates between values with $x > x_*$ and values with $x < x_*$. Clearly this fixed point is unstable, as expected based on our criteria.

Now look at a case with $(df/dx)_{x=x_*} > 1$. We use $(df/dx)_{x=x_*} = 1.5$ and again start our sequence at $x_0 = 0.001$. Figure 6.51 shows the result.

Here the mapping also carries the sequence of points away from the fixed point, but in this case the motion is in one direction with the values becoming successively greater than x_*. The fixed point is unstable, as expected. These first two examples illustrate that there are really two different types of unstable fixed points. In the case $(df/dx)_{x=x_*} < -1$ the staircase plot spirals away from the fixed point (indicating an

---

[5]Recall from the previous section that these graphs are not produced inside a *wxMaxima* session. Also, as before, we show the commands for the first example only. The remaining commands are essentially the same as the first, except that the slope of the function will be different and `load(dynamics)` need not be repeated.

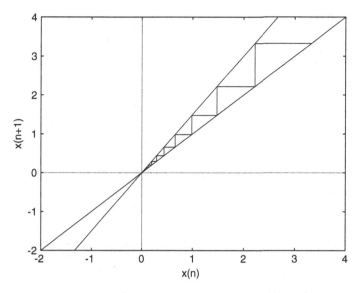

Fig. 6.51 Staircase plot for the case $(df/dx)_{x=x_*} = 1.5$ with $x_0 = 0.001$

alternating sequence). When $(df/dx)_{x=x_*} > 1$, in contrast, the staircase plot moves away from the fixed point along a single direction, bouncing between $y = f(x)$ and $y = x$.

What happens when $-1 < (df/dx)_{x=x_*} < 0$? We will investigate this case by considering the linear map with $(df/dx)_{x=x_*} = -0.8$. This time, we start a little farther away from the fixed point, at $x_0 = 0.5$. The staircase plot for this case, shown in Fig. 6.52, spirals inward toward the fixed point. This fixed point is stable, as expected. The sequence of numbers produced by the mapping converges to x_*, but with the numbers alternating between values greater than x_* and values less than x_*.

Finally, we will consider a case with $0 < (df/dx)_{x=x_*} < 1$. We examine the map with $(df/dx)_{x=x_*} = 0.8$ starting at $x_0 = 0.5$. As Fig. 6.53 illustrates, the staircase plot bounces between $y = f(x)$ and $y = x$ as it heads toward the fixed point. This fixed point is stable, as expected. The numbers in the sequence gradually approach the fixed point from above.

A little geometrical reasoning, guided by the examples given above, shows that the criteria for fixed point stability given at the beginning of this section are correct.

6.5.2 Fixed Points of the Logistic Map

We now examine the fixed points of the logistic map. First we identify the fixed points. *Maxima* solves the equation $f(x) = x$ where $f(x)$ is the logistic map function.

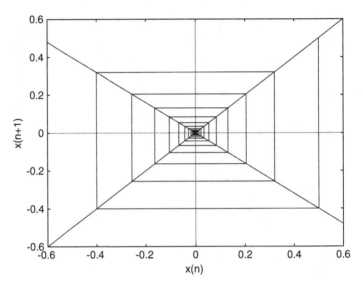

Fig. 6.52 Staircase plot for the case $(df/dx)_{x=x_*} = -0.8$ with $x_0 = 0.5$

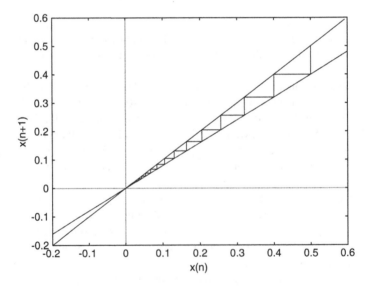

Fig. 6.53 Staircase plot for the case $(df/dx)_{x=x_*} = 0.8$ with $x_0 = 0.5$

```
(%i)  f(x)  :=  r*x*(1-x)$          fp:solve(f(x) = x, x);
(%o)  [x = r−1/r , x = 0]
```

The solution shows that the logistic map has two different fixed points: $x_1 = 1 - 1/r$ and $x_2 = 0$. We now analyze the stability of these fixed points. We evaluate the derivative of our mapping function. We define a new function for that purpose.

Then we can evaluate the derivative of our mapping function at the locations of the fixed points by substituting the solutions stored in the array fp (defined above) into the function $df(x)$.

```
(%i) df(x):=''(diff(f(x),x));
     ratsimp(df(rhs(fp[1]))); ratsimp(df(rhs(fp[2])));
(%o) df(x):= r(1-x)-rx    (%o) 2-r    (%o) r
```

The stability criteria imply that the fixed point at $x_1 = 1 - 1/r$ will be unstable for $r < 1$ and for $r > 3$, but stable for $1 < r < 3$. (In fact, for $r < 1$ this fixed point is not even within the domain of our map!) The fixed point at $x_2 = 0$ will be stable for $r < 1$ and unstable for $r > 1$.

Compare these results with the bifurcation diagrams we generated in Sect. 6.4. The bifurcation diagram shows us that for $r < 1$ the map converges to the fixed point at $x = 0$. That makes sense because we found that this fixed point is stable for those values of r, while the other fixed point is unstable (and not even in the domain). For $1 < r < 3$ the bifurcation diagram shows that the map converges to a nonzero value. It's not hard to see that this value is just our other fixed point, $x = 1 - 1/r$. Again, this behavior makes sense because for these values of r the fixed point at $x = 0$ is unstable while the one at $x = 1 - 1/r$ is stable. For $r > 3$ the bifurcation diagram reveals that the map does not converge to any single point. There is no stable fixed point for these values of r.

6.5.3 Stability of Periodic Points

So what happens in the logistic map for $r > 3$? The bifurcation diagram shows us that for values of r slightly greater than 3 the logistic map converges to a 2-cycle. We can determine the x values that make up this 2-cycle for a given value of r, and we can analyze the stability of this 2-cycle. To do so we define a new function, $f^{(2)}(x)$, that carries out two successive iterations of the map at once. In other words, $f^{(2)}(x)$ is just the composition of $f(x)$ with itself:

$$f^{(2)}(x) = f(f(x)). \tag{6.25}$$

The advantage of defining this new function is that a 2-cycle for the function $f(x)$ will be a fixed point for the function $f^{(2)}(x)$. To find the 2-cycle of a map f, and to analyze its stability, we need only to find and analyze the fixed points of $f^{(2)}$. We can find the fixed points by solving the equation $f^{(2)}(x) = x$.

```
(%i) f2(x):=f(f(x))$        soln:solve(f2(x)=x,x);
(%o) [x = -√(r²-2r-3)-r-1/2r,  x = √(r²-2r-3)+r+1/2r,  x = r-1/r,  x = 0]
```

We find that four points are fixed points of $f^{(2)}(x)$. This does not mean, however, that we have two different 2-cycles (each with two points). Two of these points are just the fixed points of f that we found earlier. A little thought reveals that the fixed

points of a map f are also solutions to the equation $f(f(x)) = x$. Thus, our fixed
points will always appear when we solve for the points in the 2-cycle. The two *new*
points are really the points in our 2-cycle. We can analyze the stability of this 2-cycle
using the same criteria we used to analyze the stability of the fixed points, but this
time using $f^{(2)}$ in place of f. (We will apply this analysis to the fixed points as well,
just to show that the results are consistent with what we found earlier.)

```
(%i) df2(x) := " (diff(f2(x), x))$
      radcan(df2(rhs(soln[1])));
      radcan(df2(rhs(soln[2])));
      radcan(df2(rhs(soln[3])));
      radcan(df2(rhs(soln[4])));
(%o) -r² + 2r + 4   (%o) - r² + 2r + 4   (%o) r² - 4r + 4   (%o21) r²
```

Both of the two points in the 2-cycle have $df^{(2)}/dx = -r^2 + 2r + 4$. This is not
surprising: both points are part of the same 2-cycle. Figure 6.54 shows a plot of
$df^{(2)}/dx$ for these 2-cycle points as well as each fixed point. The dotted lines at
$y = \pm 1$ indicate the region of stability.

The (thick gray) curve for the fixed point at $x = 0$ is within the stable region for
$r < 1$. At $r = 1$ the curve for $x = 0$ leaves the stable region, but the (thick black)
curve for $r = 1 - 1/r$ enters the stable region at that same point. This curve remains
within the stable region until $r = 3$, at which point it leaves the stable region and
the (thin black) curve for the 2-cycle enters that region. These results fit with what
we found earlier.

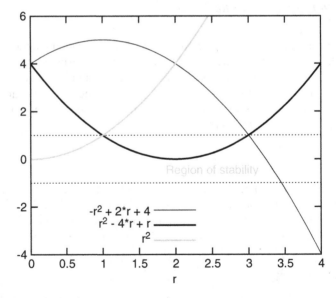

Fig. 6.54 Stability regimes for the 2-cycle and fixed points of the logistic map. The *curves* show
$(df^{(2)}/dx)_{x=x_*}$ for the 2-cycle (*thin black*), the fixed point at $x = 1 - 1/r$ (*thick black*), and the
fixed point at $x = 0$ (*thick gray*)

A question remains: at what value of r does the 2-cycle become unstable? *Maxima* can help us find out. We must find the value of r for which $df^{(2)}/dx$, evaluated at either point in the 2-cycle, is equal to -1.

```
(%i) [solve(-r^2+2*r+4=-1,r),
        float(solve(-r^2+2*r+4=-1,r))];
```
$(\%o)$ $[[r = 1 - \sqrt{6}, \ r = \sqrt{6} + 1], \ [r = -1.4495, \ r = 3.4495]]$

This equation has two solutions, but one is negative and makes no sense in our context. Therefore, we conclude that the 2-cycle becomes unstable at $r = 1 + \sqrt{6} \approx 3.4494$. What happens at this point? You might guess that there is a 4-cycle that becomes stable at this value of r. We can find and analyze this 4-cycle using the same procedure we used for the 2-cycle, but this time employing the function $f^{(4)}(x) = f(f(f(f(x))))$. Unfortunately, this process generates polynomial equations of a very high order which we may not be able to solve analytically. We could, however, solve this equation using numerical methods for a specific value of r.

Determining the values of r at which each bifurcation occurs can help us illustrate another intriguing feature of the logistic map. If the nth bifurcation occurs at $r = \gamma_n$, then we find that

$$\lim_{n \to \infty} \frac{\gamma_n - \gamma_{n-1}}{\gamma_{n+1} - \gamma_n} = \delta \qquad (6.26)$$

where $\delta \approx 4.6692$ is a number known as the *Feigenbaum number* (after Mitchell Feigenbaum, who first discovered this property). Note that the Feigenbaum relation holds exactly only in the limit $n \to \infty$, but it will be approximately correct for finite values of n.

We have found that $\gamma_1 = 3$, $\gamma_2 = 1 + \sqrt{6}$. Using the Feigenbaum relation (but ignoring the limit) we find that $\gamma_3 \approx 3.54576$. Examination of the bifurcation diagram shows that the 4-cycle does bifurcate to form an 8-cycle near this value of r.

The most remarkable aspect of the Feigenbaum relation is that it is not limited to the logistic map. In fact, this relation has been found to hold for a wide variety of dynamical systems that undergo period-doubling as some parameter is varied. This is an example of *universality*: a property of the dynamics that is largely independent of the details of the system, but rather holds for many different systems.

6.5.4 Graphical Analysis of Fixed Points

Another way to approach the analysis of the fixed points and periodic points of an iterated map is to do so graphically. The fixed points of the map f can be found by plotting $y = f(x)$ and $y = x$ and finding the points of intersection. The stability of each fixed point can be judged by estimating the slope (derivative) of $f(x)$ at each

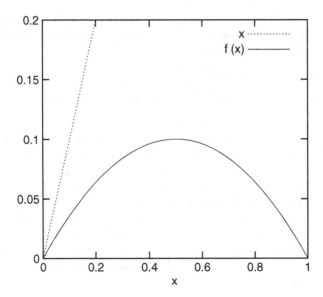

Fig. 6.55 The logistic map $f(x)$ for $r = 0.4$ (*solid*) and the line $y = x$ (*dotted*)

intersection point and applying the criteria given above. For example, we can find and analyze the fixed points of the logistic map for $r = 0.4$ by creating Fig. 6.55.[6]

```
(%i) r:0.4$ f(x) := r*x*(1-x)$
     wxdraw2d(xlabel = "x",key = "x", line_type=dots,
        explicit(x,x,0,1), line_type = solid,key=
        "f(x)",explicit(f(x),x,0,1),yrange=[0,0.2])$
```

From this plot it is clear that the only fixed point within the domain is the one at $x = 0$. It is also clear that the slope of the curve at $x = 0$ is positive, but less than 1. Therefore, we would expect the fixed point at $x = 0$ to be stable, in agreement with our earlier results for $0 < r < 1$. Now we can take a look at the same plot for $r = 2.3$, as shown in Fig. 6.56.

This plot shows that there are two fixed points in the domain, one at $x = 0$ and another at $x \approx 0.57$ (actually at $x = 1 - 1/2.3 \approx 0.5652$). The fixed point at $x = 0$ is unstable because the slope of the mapping function at that point is greater than 1. The slope at the other fixed point is negative, but still greater than -1, so this fixed point is stable. For $r > 3$ we would find that neither fixed point is stable.

We can take the same approach to examining 2-cycles by replacing $f(x)$ with $f^{(2)}(x)$. Consider the plot of $f^{(2)}(x)$ for $r = 2.8$, shown in Fig. 6.57.

---

[6]The commands for the next five graphs are much the same, so the commands that generate the second through fifth graphs are not reported.

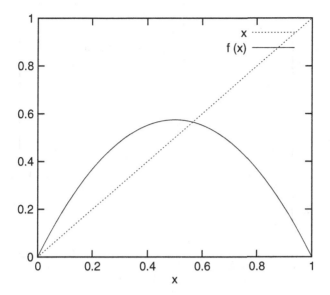

Fig. 6.56 The logistic map $f(x)$ for $r = 2.3$ (*solid*) and the line $y = x$ (*dotted*).

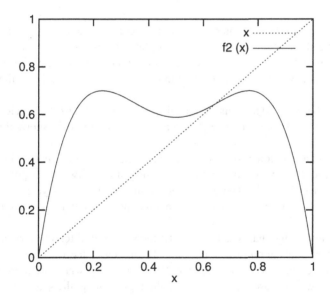

Fig. 6.57 The composite mapping function $f^{(2)}(x)$ for $r = 2.8$ (*solid*) and the line $y = x$ (*dotted*)

There are two points of intersection, but these are just our fixed points. There are no 2-cycle points. In fact, if we look back at our expressions for the coordinates of our 2-cycle we will find that they are complex-valued. We now look at the same plot for $r = 3.2$, in Fig. 6.58.

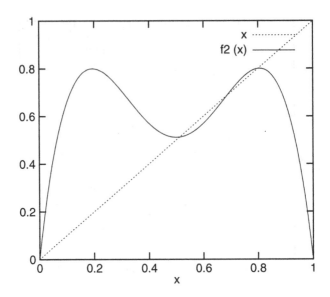

Fig. 6.58 The composite mapping function $f^{(2)}(x)$ for $r = 3.2$ (*solid*) and the line $y = x$ (*dotted*)

Now we have two new intersection points, corresponding to the points in our 2-cycle. The slope of the curve at both of these points is positive but less than 1, so the 2-cycle is stable. For $r > 1 + \sqrt{6}$ we would find that the slope of the curve at the 2-cycle points is less than -1, indicating that the 2-cycle is unstable for these values of r.

It is not hard to extend this graphical analysis to higher-order cycles. We can examine 4-cycles by plotting $y = f^{(4)}(x)$ and $y = x$. Figure 6.59 shows this plot for $r = 3.5$.

A close examination reveals eight different points of intersection in this plot. Four of these we have encountered before: they are the two fixed points and the two points of the 2-cycle. The other four are the points in a 4-cycle. The function $f^{(4)}(x)$ has a very shallow slope at these points, indicating that the 4-cycle is stable for $r = 3.5$.

Whenever the map has a stable n-cycle, the system will tend to converge to this n-cycle after many iterations of the map. In this case, the dynamics will not be chaotic. The dynamics of the map can be chaotic only when there are *no* stable n-cycles. As we increase the value of r in the logistic map, the stable fixed point at zero gives way to a stable fixed point at $x = 1 - 1/r$. This fixed point gives way to a stable 2-cycle, which gives way to a stable 4-cycle, which gives way to a stable 8-cycle, etc.

The Feigenbaum relation shows us that these period-doubling bifurcations occur at values of r that are more closely spaced as we go through the sequence. As a consequence, there is some finite value of r which exceeds the limits of the period-doubling sequence. At this value of r there are no longer any stable fixed points

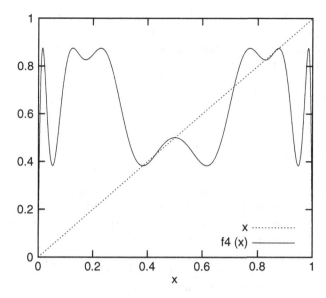

Fig. 6.59 The function $f^{(4)}(x)$ for $r = 3.5$ (*solid*) and the line $y = x$ (*dotted*)

or periodic points and the dynamics of the map is chaotic. If we use the first two bifurcation points ($\gamma_1 = 3$ and $\gamma_2 = 1 + \sqrt{6}$), and assume that the Feigenbaum relation (Eq. 6.26) holds exactly rather than only in the limit $n \to \infty$, then the properties of geometric series will show that the period-doubling sequence reaches its limit at $r \approx 3.572$. This estimate is in good agreement with the point where chaos begins in the bifurcation diagram in Fig. 6.44, as well as the point where the Lyapunov exponent becomes positive in Fig. 6.49.

There are "routes to chaos" other than the period-doubling sequence exhibited by the logistic map (and the driven damped pendulum), but the period-doubling route is a fairly typical one. Our goal here was simply to illustrate how *Maxima* can be used to explore nonlinear dynamics and chaos. For those who are interested in learning more about this fascinating part of classical mechanics we provide a list of useful references below.

- G. L. Baker and J. P. Gollub, *Chaotic Dynamics: An Introduction*, second edition, Cambridge University Press (1995).
- Robert C. Hilborn, *Chaos and Nonlinear Dynamics*, Oxford University Press (1994).
- Francis C. Moon, *Chaotic and Fractal Dynamics, An Introduction for Applied Scientists and Engineers*, Wiley (1992).
- Edward Ott, *Chaos in Dynamical Systems*, Cambridge University Press (1993).
- Steven Strogatz, *Nonlinear Dynamics and Chaos with Applications to Physics, Biology, Chemistry and Engineering*, Addison-Wesley (1994).

- Robert C. Hilborn and Nicholas B. Tufillaro, "Resource letter: ND-1: Nonlinear dynamics," *American Journal of Physics* **65**, 822–834 (1997).
- Robert DeSerio, "Chaotic pendulum: The complete attractor," *American Journal of Physics* **71**, 250–257 (2003).
- Todd Timberlake, "A computational approach to teaching conservative chaos," *American Journal of Physics* **72**, 1002–1007 (2004).

6.6   Exercises

1. Use `drawdf` to generate difference field plots (with two sample trajectories each) for the van der Pol oscillator with $\gamma = 0.7$, 1.5, and 2.5. Comment on how changing the value of γ alters the difference field and the trajectories.
2. Examine the motion of the driven van der Pol oscillator with driving frequency $\omega = 1.5$ and $\gamma = 1$. Create plots of the phase space trajectory, as well as strobe plots, for the following values of α: 0.01, 0.1, 0.2, 0.5, 0.7, 0.9, and 1.5. Comment on how the motion changes as you increase α.
3. Continue your exploration of the behavior of the driven damped pendulum for more parameter values. Keep $\omega = 2\pi$, $\omega_0 = 1.5\omega$, and $\beta = \omega_0/4$. Always launch the pendulum from rest at $\theta = -\pi/2$. Explore the motion for the values of γ shown in the table below. (Note that $\gamma = 0.2$ and 0.7 were already discussed in the text, but they are included here for completeness.) In each case, indicate whether the motion is periodic or not. If it is periodic, indicate the period (e.g., 1 if it has the same period as the driving torque, 2 if the period is twice that of the driving torque, etc.). Indicate the type of motion: either oscillatory (pendulum swings back and forth) or rolling (pendulum swings all the way around).

γ	Periodic?	Period	Type of Motion
0.7			
1.075			
1.081			
1.0824			
1.1			
1.13			
1.2			
1.35			
1.44			
1.5			

4. Figure 6.33 shows that two driven pendulums ($\gamma = 1.5$) with nearly identical initial conditions (only 0.001 radians apart) diverge noticeably at about $t = 11$ s. Modify the code used to produce this figure so that the two pendulums start only 0.0001 radians apart. At what time do these pendulums begin to noticeably

diverge? Repeat for an initial separation of 0.00001 radians. Use your results, along with Eq. 6.13, to estimate the Lyapunov exponent for the driven damped pendulum with $\gamma = 1.5$. Also estimate the Lyapunov exponent by estimating the slope of $\log D$ versus t in Fig. 6.34. Compare your two answers.

5. Create staircase plots for the logistic map with $r = 3.52$, 3.56, and 3.568. You can use $x_0 = 0.2$ in each case. Instead of using the staircase command from the dynamics package, use the code given in the text. Modify the code so that it produces 200 iterations of the map, but only plots the last 100 resulting values (ignoring the first 100). Explain what is happening as r increases from 3.1 (shown in Fig. 6.41) to 3.568. How does this behavior relate to the features seen in the bifurcation diagram (Figs. 6.43 and 6.44)?

6. Calculate the Lyapunov exponent for each of the two fixed points for the logistic map. Recall that the equation for calculating the Lyapunov exponent (Eq. 6.24) simplifies considerably if the initial point is a fixed point. Comment on how your results for the Lyapunov exponents fit with the analysis of the fixed points discussed in the text.

7. In this problem you will explore the dynamics generated by the "tent map," which is defined as

$$f(x) = r(1 - 2|x - 0.5|), \tag{6.27}$$

where $0 < r < 1$ and $0 \le x \le 1$.

(a) Make a plot of $f(x)$ for $r = 0.3$ and $r = 0.7$ to get an idea of what the tent map function looks like and how the function changes as you change the value of r.

(b) Examine staircase plots for $r = 0.3$, 0.62, 0.75, and 0.9. Use $x_0 = 0.2$ as your starting value. Describe what you see for each case. Does the sequence converge to a stable attractor? If so, what is the period? Does the sequence appear chaotic?

(c) Find the fixed points of the tent map. (Hint: there are two fixed points. You may want to rewrite the mapping function as a piecewise function before trying to solve for the fixed points. Then it is easy to find the fixed points by hand.) For each fixed point, state whether that fixed point exists for all values of r or for only a limited range of r values.

(d) Analyze the stability of each fixed point. For what range of r values (with $0 < r < 1$) is each fixed point stable? For what range of values is each fixed point unstable? Is the stability of either fixed point indeterminate for some value of r? (Note: for each fixed point you should only consider values of r for which that fixed point exists, based on your results from the previous question.)

(e) Pick a value of r for which the tent map has a stable fixed point. Examine the behavior of two trajectories that start off far apart (difference of at least 0.3). Do these trajectories converge over time?

(f) Now examine two trajectories that start off close together (difference of 0.0001), but with a value of r for which the tent map appears to generate chaotic dynamics (based on your staircase plots). Provide detailed evidence that these trajectories diverge exponentially and use your results to estimate the Lyapunov exponent for this case.

(g) Show that the Lyapunov exponent for the tent map is the same for any initial value of x. Determine an exact formula for the Lyapunov exponent, λ, as a function of r. Do this by hand! (Hint: you may want to rewrite the tent map function as a piecewise function, instead of using the absolute value. This will make it easier to take the derivative.) Compare your result to the estimate for the Lyapunov exponent that you obtained in the previous part.

(h) Make a plot of λ as a function of r. For what value of r does the Lyapunov exponent become positive?

(i) Generate the bifurcation diagram for the tent map, with $0 < r < 1$.

(j) Discuss the connections between your results for the staircase plots, the stability of the fixed points, the Lyapunov exponent, and the bifurcation diagram. How do all of these results fit together to provide a coherent picture of the transition to chaos in the tent map?

(k) Plot $f(x)$ for $r = 0.5$ and the line $y = x$ on the same plot. What is unusual about the tent map for this value of r?

(l) Compare the dynamics of the tent map, and how it changes as you vary r, to the dynamics of the logistic map and how it changes with r. In what ways are the dynamics similar? In what ways are the dynamics different? Do the two systems undergo similar dynamical changes as r is varied? If not, how are their changes different?

Erratum

Classical Mechanics with *Maxima*

Todd Keene Timberlake and J. Wilson Mixon, Jr.

© Todd Keene Timberlake & J. Wilson Mixon, Jr. 2016
T.K. Timberlake, J.W. Mixon, *Classical Mechanics with Maxima*, Undergraduate Lecture Notes in Physics, DOI 10.1007/978-1-4939-3207-8

DOI 10.1007/978-1-4939-3207-8_7

The original version of this article was inadvertently published with a sign error in equations 6.5 and 6.7. The x^2 term in both equations should have a − (negative) rather than a + (positive) in front of it.
The correct equation is shown below

$$\dot{v} = -x + \gamma(1 - x^2 - v^2)v,$$

(6.5)

$$\dot{v} = -x + \gamma(1 - x^2 - v^2)v + \alpha\cos(\omega t),$$

(6.7)

In the 15th line of page 193, $r = 0$ should be $r = 0.5$.

The online version of the original chapter can be found under
http://dx.doi.org/10.1007/978-1-4939-3207-8_6

© Todd Keene Timberlake & J. Wilson Mixon, Jr. 2016
T.K. Timberlake, J.W. Mixon, *Classical Mechanics with Maxima*, Undergraduate Lecture Notes in Physics, DOI 10.1007/978-1-4939-3207-8_7

Appendix A
Numerical Methods

A.1 The Bisection Method

In Sect. 5.3 we examined the Newton–Raphson method for numerically finding a root of a function. The Newton–Raphson method is very effective and fast, provided that the initial guess for the solution is within the basin of attraction of the iterated map used in the method. There is, however, no way to be certain that any given initial guess will be in the basin of attraction. For this reason, the Newton–Raphson method can sometimes fail to find the desired root of the function.

In this section we examine the *bisection method*, a numerical root finding method that avoids the basin of attraction problem because it does not use an iterated function. We want to use the bisection method to find a root, x_*, of the function $f(x)$ (so $f(x_*) = 0$). To use the bisection method, begin by specifying an interval $[x_L, x_R]$ which contains a single root x_* but no other roots of $f(x)$, so $x_L < x_* < x_R$. The function crosses the x-axis once and only once on this interval (at $x = x_*$), so the signs of $f(x_L)$ and $f(x_R)$ must be opposite.

The method proceeds by first computing the midpoint of this interval: $x_M = (x_L + x_R)/2$. The sign of $f(x_M)$ must be the same as either $f(x_L)$ or $f(x_R)$, but not both. We define a new interval with x_M as an endpoint and such that the function has opposite signs at the two endpoints. If $f(x_M)$ has the same sign as $f(x_L)$ (and therefore the opposite sign from $f(x_R)$) then the new interval is $[x_M, x_R]$. The root must be contained in this new, smaller interval. Otherwise the root must be contained in the interval $[x_L, x_M]$. Using our new interval, we repeat these steps (finding the midpoint, checking the signs, defining a new interval) and continue the process until the interval length is less than the desired precision for the root.

© Todd Keene Timberlake & J. Wilson Mixon, Jr. 2016
T.K. Timberlake, J.W. Mixon, *Classical Mechanics with Maxima*, Undergraduate
Lecture Notes in Physics, DOI 10.1007/978-1-4939-3207-8

The major advantage of this procedure is that for continuous functions it will always converge to a root as long as the initial interval contains a root.[1] If the initial interval contains multiple roots then the bisection method will converge to one of these roots, though perhaps not the desired one. Problems can occur if the function has a vertical asymptote or other discontinuity within the initial interval. The method can also run into difficulty with double (or quadruple, etc.) roots. Otherwise this method is quite reliable.

It might seem that we would always want to use this method rather than the Newton–Raphson method, which sometimes fails to converge. Looking at the bisection method in action, however, reveals why this is not always the case. The code below implements the bisection method for finding the root of $f(x) = \tan(x) - x - 5$ (using fpprintprec:8). We found in Sect. 5.3 that this function has a root at $x \approx 1.416$. To locate the root using the bisection method we choose the initial interval $[1.3, 1.5]$, which contains this root but no other roots of the function. The code first checks to make sure the function changes sign on the specified interval, then it implements the method, printing the number of iterations performed and the midpoint of the interval on each pass through the loop.

```
(%i) f(x) := tan(x) - x- 5$
     xL:1.3$ xR:1.5$ tol:0.00001$ n:0$
     if (f(xL)*f(xR) > 0) then
       print(
         "Sign does not change within the interval.")
       else
         for i:1 while ((i<100) and (abs(xR-xL)>tol))
           do (xM:(xL+xR)/2,
             if (f(xM)*f(xR) > 0) then xR:xM
           else xL:xM, n:n+1,
             print(n, xM ) )$
(%o) 1 1.4
     2 1.45
     3 1.425
     4 1.4125
     5 1.41875
     6 1.415625
     7 1.4171875
     8 1.4164062
     9 1.4160156
    10 1.4162109
    11 1.4161133
    12 1.4161621
    13 1.4161865
    14 1.4161743
    15 1.4161804
```

---

[1] If the function is discontinuous, it is possible for the bisection method to converge to a discontinuity rather than a root.

The bisection method does converge to the same value that the Newton–Raphson method produced, but it takes the bisection method 15 iterations to converge. The Newton–Raphson method requires four iterations (with an initial value of 1.5 and an error tolerance of 0.00001). While the bisection method is reliable, it is not very efficient.

Modify the code above to use an initial interval of $[1.5, 2]$. The function has no root on this interval, but it does have a vertical asymptote (at $x = \pi/2$) and the bisection method converges to the location of this asymptote. Try the interval $[0, 4.65]$. In that case the function has two roots and a vertical asymptote on the interval. It is hard to know to which of these values the bisection method will converge, but it must converge to one of them.

The Newton–Raphson and bisection methods illustrate a common feature of numerical algorithms: there is usually no single best algorithm. Sometimes one algorithm works better, sometimes another algorithm works better. Understanding the strengths and weaknesses of each algorithm helps us to choose the one that will work the best for the problem at hand. *Maxima*'s built-in `find_root` command uses a combination of the bisection method and a linear interpolation method that is similar to the Newton–Raphson method (but doesn't require knowledge of the derivative of the function). This combination is, on average, faster than the bisection method but more reliable than the Newton–Raphson method. The `find_root` command does give the same result as the bisection method used above.

(%i) **find_root(f(x), 1.3, 1.5);**   (%o) **1.4161843**

The difference between the final value from the bisection method and the value returned by `find_root` is approximately 4×10^{-6}, which is smaller than the 10^{-5} error tolerance that we used for the bisection method. Modifying our bisection method code to use a stricter error tolerance will produce a result that is closer to the result from `find_root`.

A.2 Numerical Integration

Maxima's library of integrals is quite large.[2] Even, so, it cannot evaluate all integrals for which closed forms exist. Furthermore, not all functions are susceptible to analytical integration.

Fortunately, *numerical* techniques allow us to compute definite integrals of functions that cannot be integrated analytically. This section introduces some examples of numerical integration, ranging from simple approximations to sophisticated methods that are built into *Maxima*.

---

[2]It incorporates material from Milton Abramowitz and Irene A. Stegun (eds.), *Handbook of Mathematical Functions*, National Bureau of Standards, U. S. Government Printing Office, 1964.

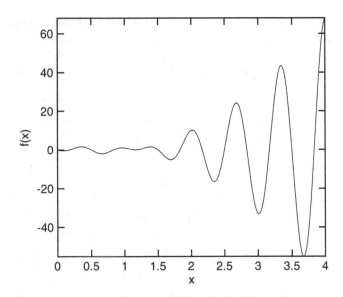

Fig. A.1 The oscillatory function $f(x) = (6x^2 - 7x)\cos(3\pi x)$

We illustrate these techniques by applying them to two functions that can be integrated analytically. Doing so provides a basis for comparing the techniques. We start with the relatively difficult case of the function

$$f(x) = (6x^2 - 7x)\cos(3\pi x). \tag{A.1}$$

As you might guess from the cosine term, this function oscillates. Figure A.1 shows the function on the interval $[0, 4]$.

```
(%i) f(x):=(6*x^2-7*x)*cos(3*%pi*x)$
     wxdraw2d(xlabel="x",ylabel="f(x)",
     explicit(f(x),x,0,4))$
```

For purposes of comparison, we use the `integrate` command to integrate the function over this interval. *Maxima* returns an exact answer, which we then evaluate as a floating-point number. This value will be compared to the results of the numerical techniques that are developed in the rest of the section.

```
(%i) integrate(f(x),x,0,4);        float(%);
(%o) 16/(3π²)(%o)    0.5403796460924681
```

Now we define a second function

$$g(x) = (x - 3)(x - 5)x^2, \tag{A.2}$$

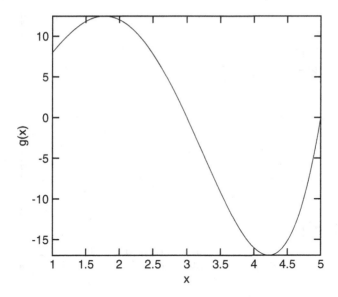

Fig. A.2 The polynomial function $g(x) = (x - 3)(x - 5)x^2$

a fourth degree polynomial. This function will be much easier to integrate with numerical methods. The function is defined and graphed over the interval $[1, 5]$ in Fig. A.2.

```
(%i) g(x):=(x-3)*(x-5)*x^2$
     wxdraw2d(xlabel="x",ylabel="g(x)",
       explicit(g(x),x,1,5))$
```

As before `integrate` provides an exact answer, and `float` provides a floating-point representation of the answer. Again, this exact answer provides the benchmark for evaluating the numerical methods that follow.

```
(%i) integrate(g(x),x,1,5);           float(%);
(%o) -16/5        (%o) -3.2
```

A.2.1 Rectangular Approximation

Our first and simplest numerical integration method breaks up the interval into equal-sized subintervals. For each subinterval we estimate the area under the curve in that subinterval by constructing a rectangle. The rectangle's width is the length of the subinterval. The rectangle's height is the value of the function at the beginning of the subinterval. The subinterval's area is simply the product of its width and height. If the function's value is negative at the beginning of the subinterval, then the area will be negative (which is the way we want it to work for integration). We then sum the areas of the rectangles for all the subintervals in order to estimate the

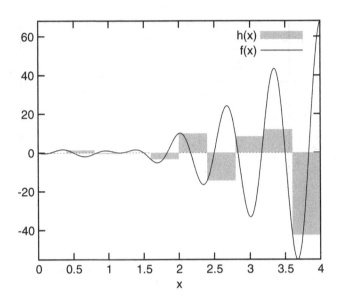

Fig. A.3 Rectangular approximation for $f(x)$

integral's value over the specified interval. This procedure is called the *rectangular approximation* for the integral. To illustrate the rectangular approximation we can define a function that gives the height of the rectangles for ten subintervals in the integration of our oscillatory function. A plot of this function, in Fig. A.3, illustrates how we will use the areas of these ten rectangles to approximate the area under the oscillatory curve.

```
(%i) h(x):= (if x <= 0.4 then f(0) else if
      x<0.8 then f(0.4) else if x<1.2 then f(0.8)
      else if x < 1.6 then f(1.2) else if x < 2
      then f(1.6) else if x < 2.4 then f(2)
      else if x < 2.8 then f(2.4) else if x < 3.2
      then f(2.8) else if x < 3.6 then f(3.2)
      else if x<4 then f(3.6) else if x >=4 then 0)$
   wxdraw2d(xaxis = true, filled_func = 0,
      fill_color= gray, key="h(x)",xlabel="x",
      explicit(h(x),x,0,4),filled_func=false,
      key="f(x)",explicit(f(x),x,0,4))$
```

The rectangles in Fig. A.3 appear as gray bars, and the curve shows the function being integrated. Notice the use of `filled_func` for showing the rectangles.[3] Cursory visual analysis shows that the rectangles do not approximate the area under

---

[3]The option `filled_func=0`, causes draw to see 0 as a function of *x*. It then shades the difference between the function $h(x)$ and 0. After this is accomplished, shading is turned off with `filled_func=false`, which is the default value.

the function very well. Even so, this analysis provides insight into the nature of numerical integration. The much more sophisticated, and accurate methods that are built in to *Maxima* are based on the same basic idea as this simple technique.

We now calculate the areas of these rectangles and sum them, using the code below. This code does not involve the $h(x)$ function defined above. That function was needed only to produce the visual representation of the rectangles in the plot above.

```
(%i) a:0$ b:4$ n:10$ dx:(b-a)/n$ int:0$
     for i:0 thru n-1 do
       block(xi:a+i*dx,int:int+f(xi)*dx)$
     float(int);
(%o) −11.477
```

This result is not good at all, giving an estimated value of -11.477 versus the actual value of 0.5404. Re-evaluate the code block above using 100 subintervals (n:100), and again with 1000 subintervals. The results get better as n gets larger, but even with n:1000, the error is still around 25 %.

Now we apply this approach to the polynomial function, $g(x)$. Again, we look at the rectangles by creating a function that gives the height of the rectangle for each subinterval and then plotting those rectangles (along with the polynomial function).[4]

Figure A.4 indicates that the rectangular approximation should work better for this function. The code below evaluates the rectangular approximation for this integral.

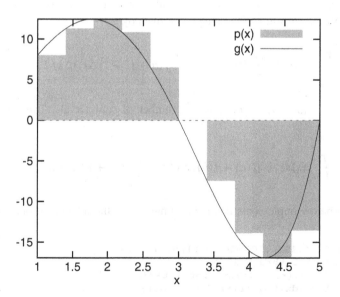

Fig. A.4 Rectangular approximation for $g(x)$

---

[4]The code is a straightforward modification of the code above and is omitted.

```
(%i)  a:1$ b:5$ n:10$ dx:(b-a)/n$ int:0$
      for i:0 thru n-1 do
      block(xi:a+i*dx, int:int+g(xi)*dx)$
        float(int);
(%o) −1.0701
```

Even for this relatively well-behaved function, with `n:10` the result (−1.07) is not very close to the actual value (−3.2). Confirm that with `n:100` the approximation (−3.035) is much better, and with `n:1000` it is even better (−3.184). With the use of modern computers, `n:10000` or more could be used, so that the rectangular approximation could be forced to yield good results, at least for reasonably well-behaved functions. We can do much better, however, by selecting a better numerical method.

A.2.2   Trapezoidal Approximation and Simpson's Rule

Instead of using rectangles, we could use trapezoids to estimate the subinterval areas. The bottom of the trapezoid will still run along the x-axis over the subinterval. The height of the left side of the trapezoid is given by the function's value at the start of the subinterval, and the height of the right side is given by the function's value at the end of the subinterval. This minor modification of the rectangular approximation yields the *trapezoidal approximation*.

The area of the trapezoid is just the width times the average of the two heights, so the approximate value of the integral is given by

$$\int_a^b f(x)dx \approx \sum_{i=0}^{n-1} (f(a+i\Delta x) + f(a+(i+1)\Delta x))\Delta x/2 \qquad (A.3)$$

where $\Delta x = (b-a)/n$ and n is the number of subintervals. This sum can be rewritten as

$$\int_a^b f(x)dx \approx (f(a) + f(b))\Delta x/2 + \sum_{i=1}^{n-1} f(a+i\Delta x)\Delta x. \qquad (A.4)$$

The code below implements this approximation for the integral of our oscillatory function $f(x)$.

```
(%i) a:0$ b:4$ n:10$ dx:(b-a)/n$ int:0$
     for i:1 thru n-1 do
       block(xi: a+i*dx, int:int+f(xi)*dx)$
     float(dx*0.5*(f(a)+f(b))+int);
(%o) 2.1227
```

The result isn't impressive for `n:10` (2.1226 versus 0.5404). It is, however, much better than the rectangular approximation. Furthermore, the quality improves quickly with increased n: for n:100 the trapezoidal approximation actually gives a pretty good result (0.5468). As an exercise, determine how close the approximation becomes with n:1000 The results for this difficult oscillatory function suggests that the use of the trapezoidal rule represents progress.

We now apply the trapezoidal method to the integral of our polynomial function $g(x)$.

```
(%i) a:1$ b:5$ n:10$ dx:(b-a)/n$ int:0$
     for i:1 thru n-1 do
       block(xi:a+i*dx,int:int+g(xi))$
     float(dx*(0.5*(g(a)+g(b))+int));
(%o) −2.6701
```

The result is not great for n:10 (−2.67 versus −3.4). Confirm that with n:100 the result is fairly accurate. Still, we can do better.

The rectangular approximation represents the original function as a constant within each subinterval. The trapezoidal approximation represents the function as a straight line within each subinterval. The trapezoidal approximation does a better job because the straight line segments more closely follow the shape of the function than can a set of flat-line (constant) segments. The straight lines are more flexible than the constant values.

Even more flexibility results from allowing the top of the shape to *curve*. Suppose we have two adjacent subintervals with endpoints x_0, x_1, and x_2. We can approximate the function on the interval $[x_0, x_2]$ by a parabola that passes through the points $(x_0, f(x_0))$, $(x_1, f(x_1))$, and $(x_2, f(x_2))$. We can then integrate this parabolic function to find the area underneath the parabola from x_0 to x_2. The result is

$$(f(x_0) + 4f(x_1) + f(x_2))\Delta x/3. \tag{A.5}$$

This serves as an approximation for the integral of our function from x_0 to x_2. Repeating this procedure for each pair of subintervals and then summing the results yield *Simpson's Rule* which gives the following approximation for the integral:

$$\int_a^b f(x)dx \approx (f(x_0) + 4f(x_1) + 2f(x_2) + 4f(x_3) + \ldots + 2f(x_{n-2}) + 4f(x_{n-1}) + f(x_n))\Delta x/3, \tag{A.6}$$

where $x_i = a + i\Delta x$, $\Delta x = (b-a)/n$, and the number n of subintervals is assumed to be even.

The code below implements Simpson's Rule for our oscillating function. Note that the if statement evaluates to 2 if the index of the point (the i in x_i) is even, and 4 if the index is odd. The last step adds the endpoint terms ($f(x_0)$ and $f(x_n)$).

```
(%i) a:0$ b:4$ n:10$ dx:(b-a)/n$ int:0$
     for i:1 thru n-1 do
       block(xi:a+i*dx,int:int+(if mod(i,2)=0
       then 2 else 4)*f(xi))$
     float(dx*(f(a)+f(b)+int)/3);
(%o) −4.5795
```

The results from Simpson's Rule are not impressive for n:10 (remember: the actual value is approximately 0.5404), but it works pretty well with n:100 and much better with n:1000. Confirm these assertions.

We now apply Simpson's rule to $g(x)$, our polynomial function.

```
(%i) a:1$ b:5$ n:10$ dx:(b-a)/n$ int:0$
     for i:1 thru n-1 do
       block(xi:a+i*dx,int:int+(if mod(i,2)=0
       then 2 else 4)*g(xi))$
       float(dx*(g(a)+g(b)+int)/3);
(%o) −3.1863
```

The result is pretty good with n:10. With n:100 we get a result that is accurate to better than one part in a million.

A.2.3   Monte Carlo Methods

Another way to estimate the value of an integral numerically is to use a so-called Monte Carlo method. These methods are so named because they are based on randomness and probability, much like the games of chance at the famous casinos in Monte Carlo, Monaco.

The Monte Carlo method that we will examine here is the *sample mean method*. It is based on the mean-value theorem of integral calculus which states that the integral of a function over an interval equals the mean value of the function on that domain times the length of the interval. We can estimate the mean value of the function by evaluating the function at several randomly selected points and taking the mean of these values. Multiplying this result by the length of the interval yields an estimate for the integral of the function. The following code applies this procedure to $f(x)$, our oscillatory function.

```
(%i) a:0$ b:4$ n:10$ dx:float(b-a)$ int:0$
     for i:1 thru n do
       block(xi:a+random(dx),int:int+f(xi))$
       float((b-a)*int/n);
(%o) 6.3576
```

Unless you got lucky (remember, random chance is involved) the result isn't very good. Simpson's rule definitely works better here. In fact, even the rectangular approximation often works better. (Remember: This approach yields random results, so a given pass might yield an estimate that is near to the actual value.) If you increase the value of *n* you will get closer to the correct answer. Even so, this method does not seem to work well for this function. Let's try this method on our polynomial function.

```
(%i) a:1$ b:5$ n:10$ dx:float(b-a)$ int:0$
     for i:1 thru n do
       block(xi:a+random(dx),int:int+g(xi))$
       float((b-a)*int/n);
(%o) 0.30915
```

This method is still not very good. We will not prove it here, but the errors from the rectangular approximation scale like n^{-1}, while the trapezoidal approximation errors scale like n^{-2} and the Simpson's Rule errors scale like n^{-4}. Errors from Monte Carlo methods, on the other hand, scale like $n^{-1/2}$. All of the classical approximation techniques considered above perform better than the Monte Carlo methods, because as n increases their errors decrease much faster (fastest for Simpson's Rule).

So why even consider Monte Carlo methods? The answer is that they are useful for higher dimensional integrals. Consider a d-dimensional integration. For the classical methods like the rectangular approximation or Simpson's Rule if the errors in a one-dimensional integral scale like n^{-k}, then the errors for a d-dimensional integral scale like $n^{-k/d}$. In contrast, Monte Carlo errors scale like $n^{-1/2}$ regardless of the dimension of the integral. Thus for a double integral the rectangular approximation errors scale like $n^{-1/2}$, the same as for Monte Carlo methods. For a triple integral Monte Carlo methods are *better* than the rectangular approximation, and for an integral of 9 or higher dimensions Monte Carlo methods perform better than Simpson's Rule. Furthermore, Monte Carlo methods are *much* easier to implement than the classical rules for high dimensional integrals.

Consider approximating the function of two variables given by

$$q(x, y) = (x^2 + y^2) \exp(-x^2 - y^2). \tag{A.7}$$

We use the code below to plot this function for $-3 < x < 3$ and $-2 < y < 2$, which will also serve as our domain of integration. In Fig. A.5, the meshed black surface represents the function. The gray plane is the $z = 0$ plane. Therefore, the volume between the surface and the plane is what we seek to determine.

```
(%i) q(x,y):=(x^2+y^2)*exp(-x^2-y^2)$
     wxdraw3d(view=[45,315], color=gray,ztics=.1,
     xlabel="x",ylabel="y",
     explicit(0,x,-4,4,y,-3,3),key="q(x,y)",
     color=black,explicit(q(x,y),x,-3,3,y,-2,2) )$
```

Fig. A.5 The two-dimensional function $q(x, y) = (x^2 + y^2) \exp(-x^2 - y^2)$

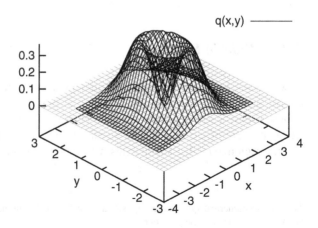

Now we will evaluate the integral of this function on the specified domain using *Maxima*'s `integrate` command. *Maxima* calculates the volume between the gray plane and the black surface in the plot above. The *Maxima* output shows the exact result and its floating-point representation.

```
(%i) integrate(integrate(q(x,y),x,-3,3),y,-2,2);
     float(%);
```

$$(\%\text{o}) \quad -e^{-9}\left(\sqrt{\pi}\,\left(2\,e^{5}\,\text{erf}\,(3) + 3\,\text{erf}\,(2)\right) - e^{9}\,\pi\,\text{erf}\,(2)\,\text{erf}\,(3)\right) \quad (\%\text{o})\ 3.0612$$

Now we can estimate the integral using Monte Carlo methods. The code appears below. The code is only slightly longer, and no more complicated, than the Monte Carlo code for the one-dimensional case.

```
(%i) x1 :-3$ x2 :  3$ y1 : -2$ y2 : 2$ n : 10$
     dx:float(x2-x1)$ dy:float(y2-y1)$ int:0$
     for i:1 thru n do
       block(xi:x1+random(dx),yi:y1+random(dy),
         int:int+q(xi,yi))$
     float((x2-x1)*(y2-y1)*int/n);
(%o) 4.7005
```

As with the one-dimensional case, when you execute this input block, the result may look pretty good or it may not. It's random! One way to get a good estimate for the error is to re-evaluate the code without changing n. The result is a different answer each time because we have not reset the state of the random number generator. The spread of the results provides an estimate for the error in the numerical integration.

Once you have run the code several times with `n:10`, change to `n:100` and evaluate the code several times. Some variation will still be present, but you should notice that the results are all hovering closer to the correct answer. Try it again with `n:1000`. For high-dimensional integrals Monte Carlo methods are more effective and easier to implement than the classical methods, and they provide their own built-in way for estimating error (by evaluating multiple times and looking at the spread and/or the standard deviation of the results).

A.2.4   Built-in Routines

Now that we know something about numerical integration we look at *Maxima*'s built-in numerical integrators. We look first at `quad_qags`.[5] This command uses a more sophisticated algorithm than the classical methods we examined above. It adapts the size of the subintervals based on how rapidly the function is changing in that region (smaller subintervals when the function changes rapidly, larger

---

[5]The term "numerical quadrature" is used as a synonym for numerical integration. Hence the "quad" in the names of the procedures.

subintervals when the function is relatively constant). We use quad_qags to compute the integral of our oscillatory function $f(x)$ from $x = 0$ to $x = 4$.

```
(%i) quad_qags(f(x),x,0,4);
(%o) [0.5403796460924575, 3.929944130537377 10⁻¹¹, 147, 0]
```

The output from quad_qags is a list with four numbers. The first number is the estimate for the integral, which is the same as the 0.54037964609246 that we obtained analytically. The second number is an estimate for the absolute error in the result. In this case, the error is *very* small compared to the result itself. (In fact, this number is an overestimate of the error. We can see by comparison with the analytical result that the actual error is 0 to all decimal places shown.) The third number shows how many times the function had to be evaluated in estimating the integral. The number, 147, might seem large, but go back and try any of the other methods with n:147 (note that each method evaluates the integrand at least n times) and compare the result to the quad_qags result. You should be impressed that quad_qags gets such a good result with so few function evaluations. The final output number is an error code that is described in the quad_qags documentation (a code of 0 indicates normal output).

Now try another built-in routine: quad_qag. This routine is specifically designed to work with oscillatory integrands. It has the same general form as quad_qags, but uses a fifth argument to specify the degree of the approximation formula used in computing the estimate. We will use degree 6.

```
(%i) quad_qag(f(x),x,0,4,6);
(%o) [0.5403796460924478, 5.33738358364152 10⁻¹³, 61, 0]
```

The quad_qag procedure yields an even better result (smaller estimated error) than quad_ qags, and it took only 61 function evaluations instead of 147. We consider one more routine, called romberg. This command uses a routine called Romberg's method, which is similar to Simpson's method, but executes faster. The required error tolerance can be set by assigning a value to the variable rombergtol (romberg tolerance) as shown below. We set this level at the same order of magnitude that results from applying quad_qag.

```
(%i) rombergtol:1e-13$ romberg(f(x),x,0,4);
(%o) 0.5403796460924576
```

The Romberg method yields a good result for this function, but it does not provide an error estimate (although presumably it is on the order of $1e - 13$ as required) and we do not find out how many times it had to evaluate the function. Generally quad_qags or quad_qag will be better for one-dimensional integrals, but romberg can be used for higher dimensional integrals. The routines in the quadpack package, which includes quad_qag and quad_qags, cannot. As an example, we can use romberg to evaluate the multiple integral of the function $q(x, y)$ that we examined above.

```
(%i) rombergtol:1e-6$
     romberg(romberg(q(x,y),x,-3,3),y,-2,2);
(%o) 3.061249076983766
```

The result is equal to the 3.0612 that we derived analytically, at least to four decimal places. The `romberg` method can be applied to higher-order nestings, too. In fact, `romberg` can even handle variables in the limits of integration as long as they only appear on "inner" integrals (i.e., integrals nested inside other integrals) and only involve the integration variables from "outer" integrals (*i.e.*, integrals within which they are nested). Below is an example in which the inner integral is over y, but has an x in its limits of integration. The outer integral is over x and has no variables in its limits of integration. In the code below we first evaluate the integral exactly, and then evaluate it using `romberg`.

```
(%i) integrate(integrate(q(x,y),y,-2,x),x,-3,3);
     float(%);
     rombergtol:1e6$
     romberg(romberg(q(x,y),y,-2,x),x,-3,3);
```
$$(\%o) \quad -\frac{e^{-9}\left(\sqrt{\pi}\left(2e^5\,\mathrm{erf}(3)+3\,\mathrm{erf}(2)\right)-e^9\,\pi\,\mathrm{erf}(2)\,\mathrm{erf}(3)\right)}{2}$$
(%o) 1.530624540495877   (%o) 1.53062454031224

So we see that `romberg` does a good job in estimating this integral.

What if we want to numerically evaluate an integral in which one or more limit of integration is infinite? Consider the following example, which can be evaluated exactly.

```
(%i) r(x) := exp(-x)*log(x) $
     integrate(r(x),x,0,inf);     float(%);
```
(%o) $-\gamma$ (%o) -0.5772156649015329

Here *Maxima* evaluates this integral exactly in terms of the Euler-Mascheroni constant γ. The floating-point representation is produced for comparison.

To numerically compute this integral we could try to change variables so that the new integral would be over a finite interval, or use other sophisticated techniques like Fourier transforms. But a straightforward way to handle the situation is to use `quad_qagi`, which is like `quad_qags` but accepts infinite limits of integration. We use `quad_qagi` to evaluate our example integral below.

```
(%i) quad_qagi(r(x),x,0,inf);
```
(%o) $[-0.5772156649015293,\ 5.110526668516968\ 10^{-9},\ 345,\ 0]$

The result is excellent, but it required many function evaluations. Also, the size of the absolute error is larger than in previous examples. This relatively large error shows that numerical integration doesn't work as well when the limits are infinite (which is not surprising). So, if possible, one should recast an infinite integral as a finite one. If that is not possible, then `quad_qagi` provides a relatively good numerical approximation. Finally, consider an example in which both limits are infinite.

```
(%i) integrate(exp(-x^2),x,minf,inf); float(%);
     quad_qagi(exp(-x^2),x,minf,inf);
```
(%o) $\sqrt{\pi}$ (%o) 1.772453850905516)
(%o) $[1.772453850905516, 1.420263678094492\ 10^{-8}, 270, 0]$

Again we see that `quad_qagi` gives a good approximation for this integral.

Each of the methods introduced in this section has its strengths and weaknesses. Because modern computers allow almost instantaneous numerical analysis of many functions, one need not limit analysis to a single, best approach. Try two or three. If the results are the same, then you probably have your answer. If not, then undertake a more careful examination to determine which method is giving the best results.

A.3 Runge–Kutta Algorithms

In Sect. 5.4 we examined two algorithms for numerically integrating a system of ordinary differential equations (ODEs): the Euler algorithm and the Euler–Cromer algorithm. In that section we found that the Euler algorithm is unstable, causing the calculated energy of the system to grow over time and leading to large errors in the solution over long times. The Euler–Cromer algorithm, on the other hand, is stable for oscillatory systems. In this section we examine a class of algorithms for numerically integrating ODEs that are known as *Runge–Kutta* algorithms. The Runge–Kutta algorithms can provide superior performance compared to the Euler and Euler–Cromer algorithms.

The Euler and Euler–Cromer algorithms discussed in Sect. 5.4 used a fixed time step of size Δt. We saw that a smaller value for Δt generally leads to a more accurate result, but with the cost of a larger number of function evaluations (and thus more time for the algorithm to run). These algorithms were derived by using a simple Taylor series to first order in Δt, so the calculation of each time step introduces errors that are proportional to Δt^2 and higher powers of Δt. We say that the errors are "of order Δt^2," which we write as $O(\Delta t^2)$. Because these errors accumulate, the *global error* (i.e., the maximum error for any part of the solution) is $O(\Delta t)$. So the smaller Δt, the smaller the global error in our numerical solution. Because the global error is of first order in the time step, we say that these two algorithms are *first order* algorithms.

We can get better performance by finding an algorithm that eliminates the $O(\Delta t^2)$ errors in the computation of each time step. If we can accomplish that goal, then our algorithm will have $O(\Delta t^3)$ errors for each time step and a $O(\Delta t^2)$ global error. Such an algorithm is considered a *second order algorithm*. There are several different second order algorithms. The one we will examine here is known as the second order Runge–Kutta algorithm. In all Runge–Kutta algorithms we evaluate the rate of change of each variable multiple times for each time step. For example, the second order Runge–Kutta algorithm requires evaluation of the rate of change twice per time step (for each variable), as compared to just once for the Euler and Euler–Cromer algorithms.

The second order Runge–Kutta algorithm, as applied to Newton's equations of motion, is fairly simple. We just use the Euler algorithm to estimate the position and velocity at the midpoint of the time step ($t = t_n + \Delta t/2$), and then use the estimated midpoint position and velocity to estimate the rates of changes (i.e., velocity and acceleration) for the time step. The second order Runge–Kutta algorithm for solving Newton's equations of motion for a single particle in one dimension are:

$$k_{x1} = v_n \Delta t,$$

$$k_{v1} = \frac{1}{m} F(x_n, v_n, t_n) \Delta t,$$

$$k_{x2} = (v_n + \frac{1}{2} k_{v1}) \Delta t,$$

$$k_{v2} = \frac{1}{m} F(x_n + \frac{1}{2} k_{x1}, v_n + \frac{1}{2} k_{v1}, t_n + \frac{1}{2} \Delta t) \Delta t,$$

$$x_{n+1} = x_n + k_{x2},$$

$$v_{n+1} = v_n + k_{v2},$$

(A.8)

where $F(x, v, t)$ is the force on the particle.

The k-notation in Eq. A.8 is a standard one for presenting the Runge–Kutta algorithms. However, we can present the second order Runge–Kutta algorithm in a simpler form by first computing the midpoint values of t, x, and v:

$$t_m = t_n + \frac{1}{2} \Delta t,$$

$$x_m = x_n + \frac{1}{2} v_n \Delta t,$$

$$v_m = v_n + \frac{1}{2m} F(x_n, v_n, t_n) \Delta t,$$

$$x_{n+1} = x_n + v_m \Delta t,$$

$$v_{n+1} = v_n + \frac{1}{m} F(x_m, v_m, t_m) \Delta t.$$

(A.9)

To demonstrate the use of this algorithm we consider the same harmonic oscillator system that we examined with the Euler and Euler–Cromer algorithms in Sect. 5.4. The code below implements the second order Runge–Kutta algorithm for the harmonic oscillator with the same parameters used in Sect. 5.4 for the other two algorithms ($k = 1$ N/m, $m = 1$ kg, $x(0) = 1$ m, $v(0) = 0$, and a time step of $\Delta t = 0.2$ s). Note that we don't bother to compute the midpoint time t_m because it is not needed.

```
(%i) k:1$ m:1$ dt:0.2$ nt:100$
     array(xrk,nt)$
     array(vrk,nt)$
     xrk[0]:1$ vrk[0]:0$
     for i:0 while i < nt do
       block(xm:xrk[i]+0.5*dt*vrk[i],
         vm:vrk[i]-0.5*dt*k*xrk[i]/m,
       xrk[i+1]:xrk[i]+vm*dt,vrk[i+1]:vrk[i]-k*xm*dt/m)$
```

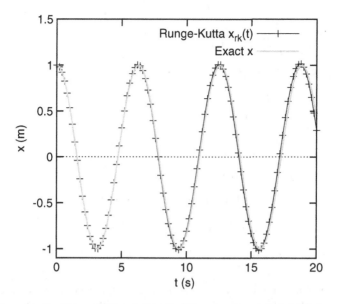

Fig. A.6 Second order Runge–Kutta solution ($x_{rk}(t)$) and exact solution ($x(t)$) for a harmonic oscillator

We can then make a plot of the position versus time data generated by the algorithm and compare it to the exact solution ($x(t) = \cos(t)$), as shown in Fig. A.6. The code used to generate this plot is very similar to the code used to generate similar plots in Sect. 5.4, so we will not show the code here.

An examination of this plot shows that the second order Runge–Kutta algorithm does a good job of approximating $x(t)$ on the interval shown. Initially the results of the algorithm are indistinguishable from the exact solution. However, by $t = 20$ s the results of the algorithm are noticeably different from the exact solution. A comparison of this plot to the plots for the Euler and Euler–Cromer algorithms in Sect. 5.4 shows that the second order Runge–Kutta algorithm performs better than the Euler–Cromer algorithm (and vastly better than the Euler algorithm) on this problem. We should expect better performance from the second order Runge–Kutta algorithm since the global error should be $O(\Delta t^2)$ rather than $O(\Delta t)$ as for the Euler–Cromer algorithm.

We can get even better performance by using a higher-order algorithm. For example, the fourth-order Runge–Kutta algorithm for numerically integrating Newton's equations of motion for a single particle in one dimension is:

$$k_{x1} = v_n \Delta t,$$

$$k_{v1} = \frac{1}{m} F(x_n, v_n, t_n) \Delta t,$$

$$k_{x2} = (v_n + \frac{1}{2} k_{v1}) \Delta t,$$

$$k_{v2} = \frac{1}{m} F(x_n + \frac{1}{2} k_{x1}, v_n + \tfrac{1}{2} k_{v1}, t_n + \frac{1}{2} \Delta t) \Delta t,$$

$$k_{x3} = (v_n + \frac{1}{2} k_{v2}) \Delta t,$$

$$k_{v3} = \frac{1}{m} F(x_n + \frac{1}{2} k_{x2}, v_n + \tfrac{1}{2} k_{v2}, t_n + \frac{1}{2} \Delta t) \Delta t, \qquad \text{(A.10)}$$

$$k_{x4} = (v_n + k_{v3}) \Delta t,$$

$$k_{v4} = \frac{1}{m} F(x_n + k_{x3}, v_n + k_{v3}, t_n + \Delta t) \Delta t,$$

$$x_{n+1} = x_n + \frac{1}{6} (k_{x1} + 2k_{x2} + 2k_{x3} + k_{x4}),$$

$$v_{n+1} = v_n + \frac{1}{6} (k_{v1} + 2k_{v2} + 2k_{v3} + k_{v4}).$$

This algorithm is obviously quite a bit more complicated than the second order Runge–Kutta algorithm or the Euler–Cromer algorithm. For example, we must evaluate the force function four times per time step in the fourth order Runge–Kutta algorithm while we only had to evaluate the force function twice for the second order Runge–Kutta algorithm. These additional function evaluations cost time, so it might seem like we could do better by just reducing the time step in our second order algorithm. The second order algorithm with a time step of $\Delta t/2$ will have about the same number of function evaluations as the fourth order algorithm with time step Δt. But the fourth order algorithm has a $O(\Delta t^4)$ global error. With a small time step, this error will usually be much smaller than the global error for the second order algorithm with a $\Delta t/2$ time step.

Fortunately we don't have to program the fourth order Runge–Kutta algorithm in *Maxima*, because the built-in rk command implements this algorithm. We used rk earlier in the text (for example, in Sect. 2.3.5). The code below shows how to use rk to numerically integrate the equations of motion for our simple harmonic oscillator example. The arguments of the rk command are lists. The first list supplies the expressions for evaluating the derivatives (in this case, dx/dt and dv/dt). The second list gives the variables to be solved (x and v). The third list gives initial conditions (values for $x(0)$ and $v(0)$). The ordering (first x, then v) *must* be the same for all three of these lists. The final list gives the range of times over which the solution is to be generated, as well as the time step (in this case we want the solution for $0 < t < 20$ with a time step of 0.2). Figure A.7 shows a plot of $x(t)$ using the results from the rk command as well as the exact solution.

```
(%i) kill(values,functions,arrays)$
     k:1$ m:1$ dt:0.2$ x0:1$ v0:0$
     sol:rk([v,-k*x/m],[x,v],[x0,v0],[t,0,20,dt])$
```

Note that the numerical solution is indistinguishable from the exact solution in this plot. This illustrates the improved performance that we expected from the fourth order Runge–Kutta algorithm, as compared to the second order algorithm. Since we

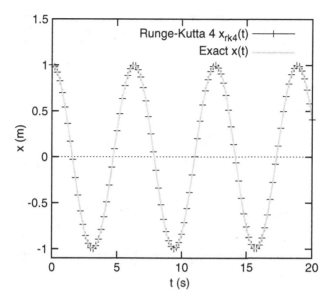

Fig. A.7 Fourth order Runge–Kutta solution $(x_{rk4}(t))$ and exact solution $(x(t))$ for a harmonic oscillator

know the exact solution we can take a closer look by computing the actual error for each x value produced by rk as a function of t. The code below generates a list of such error values at each time step and then constructs a plot of error as a function of time, as shown in Fig. A.8.

```
(%i) err:makelist([sol[i][1],sol[i][2]-cos(sol[i][1])],
       i,1,length(sol))$
     wxdraw2d(xaxis = true, user_preamble=
       "set key top left", point_size=1,
       points_joined=true, xlabel="t (s)",
       ylabel="error (m)", points(err))$
```

We see that the error oscillates, but with increasing amplitude over the time interval. The largest errors occur at the end of the interval, with the size of the error roughly 0.0002 m. Considering that the amplitude of the particle's oscillations is 1 m, this is a relatively small error.

The fourth order Runge–Kutta method is a very popular one for the numerical integration of ODEs. For many users, this algorithm may be the only one they ever need. Even better algorithms are available, however. One of the big advances in the numerical solution of ODEs was the development of adaptive step size methods. The idea is to change the size of the step used in the algorithm depending on the size of the rate of change for the variables. A smaller step size is used when the rates of change are large, but a larger step size can be used when the rates of change are small. Adapting the step size in this way provides both accuracy (because the step size can be reduced to avoid large errors) and speed (because the step size can be increased, reducing the number of steps needed, as long as accuracy won't be compromised).

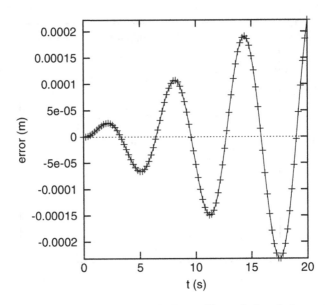

Fig. A.8 Error $(x_{rk4}(t) - x(t))$ of the fourth order Runge–Kutta solution of a harmonic oscillator

Maxima has a built-in routine that uses a modified Runge–Kutta algorithm with adaptive step size. The `rkf45` command uses the Runge–Kutta-Fehlberg method of fourth-fifth order to numerically integrate a systems of ODEs. This method uses a fourth order Runge–Kutta algorithm in tandem with a fifth order Runge–Kutta algorithm. The difference between the results of the two algorithms is used to estimate the error, and the step size can then be adjusted if errors become too large. The `rkf45` command allows the user to set the absolute error tolerance for each step by specifying a value for `absolute_tolerance` (the default value is 10^{-6}).

Below we use the `rkf45` command to numerically integrate the ODEs for our example harmonic oscillator system. The syntax is almost identical to that of `rk`. Here we use the option `report=true` in order to get some useful information about the performance of the algorithm.

```
(%i) kill(values,functions,arrays)$ load(rkf45)$
     k:1$ m:1$ dt:0.2$ x0:1$ v0:0$
     sol45: rkf45([v,-k*x/m],[x,v],[x0,v0],
     [t,0,20,dt],report=true$
```

Info: rkf45:
Integration points selected: 140
Total number of iterations: 140
Bad steps corrected: 1
Minimum estimated error: 4.438211725433255 10^{-8}
Maximum estimated error: 5.669538788115278 10^{-7}
Minimum integration step taken: 0.07957609597782067
Maximum integration step taken: 0.1529355426216121

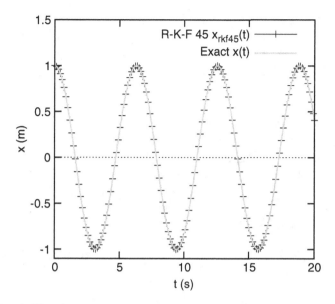

Fig. A.9   Fourth-fifth order Runge–Kutta–Fehlberg (R–K–F) solution ($x_{rkf45}(t)$) and exact solution ($x(t)$) for a harmonic oscillator

The estimated maximum error for any step is $\approx 6 \times 10^{-7}$, which is less than the default error tolerance of 10^{-6}. This level of accuracy was achieved with only 140 steps, as compared to the 100 steps used by rk in the previous example. Figure A.9 plots position as a function of time for both the rkf45 results and the exact solution.[6]

It is clear that the algorithm worked well. The numerical results are indistinguishable from the exact solution in the plot. Let's take a closer look by computing the actual error in the numerical results, which appear in Fig. A.10.

We see that the error oscillates with increasing amplitude, just as it did for the fourth order Runge–Kutta solution. However, the largest error on the interval is about 10^{-5} m, about 20 times smaller than the largest error from rk. Recall that this improved accuracy was achieved with roughly the same number of time steps (140 versus 100).

If we need more accurate results we can set absolute_tolerance to a smaller value in the rkf45 command. The code below shows the error results for our harmonic oscillator using rkf45 with an error tolerance of 10^{-7}. Figure A.11 shows the errors.

---

[6]Once again, the commands are essentially the same as those used above, and they are omitted. Likewise, for Fig. A.10.

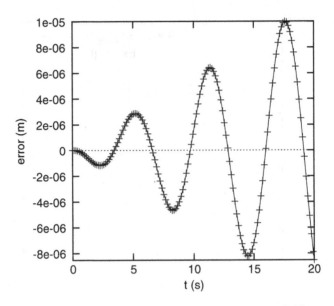

Fig. A.10 Error $(x_{rkf45}(t) - x(t))$ of the fourth-fifth order Runge–Kutta–Fehlberg solution of a harmonic oscillator

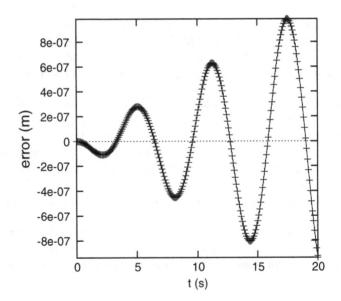

Fig. A.11 Same as Fig. A.10, but with a smaller error tolerance

```
(%i) kill(values,functions,arrays)$ load(rkf45)$
     k:1$ m:1$ dt:0.2$ x0:1$ v0:0$
     sol45b:rkf45([v,-k*x/m],[x,v],[x0,v0],
       [t,0,20,dt],absolute_tolerance=1e-7,report=true)$
     err45b:makelist([sol45b[i][1],sol45b[i][2]-
       cos(sol45b[i][1])],i,1,length(sol45b))$
     wxdraw2d(xaxis = true, user_preamble= "set key
       top left", point_size=1,points_joined=true,
       xlabel="t (s)",ylabel="error (m)", points(err45b))$
```

Info: rkf45:
Integration points selected: 248
Total number of iterations: 248
Bad steps corrected: 1
Minimum estimated error: $2.03214853050808 \ 10^{-10}$
Maximum estimated error: $5.399519048049547 \ 10^{-8}$
Minimum integration step taken: 0.02047937767783381
Maximum integration step taken: 0.08610004786735029

These results show that the rkf45 command achieved an estimated maximum error per step of $\approx 5.4 \times 10^{-8}$ and a maximum actual error of about 10^{-6} with only 248 steps. The performance of rkf45 will generally be superior to that of rk, and particularly so if the rates of change of the variables vary greatly over the integration interval.

There are even more sophisticated algorithms for numerically integrating systems of ODEs, but currently the only built-in routines that are provided in *Maxima* are rk and rkf45. For most users these two routines will suffice.

A.4 Modeling Data

Physics is about the real world, not just the theories we have devised to describe and explain that world. Therefore, physicists must work with actual experimental data. In this section we examine some *Maxima* tools that can be used to build a mathematical model to represent a set of data. We will focus on cases for which the mathematical model is just a function of a single variable. Once the function has been obtained, then it can be manipulated or displayed using the wide variety of *Maxima* commands discussed in this book.

We examine two general methods of devising a function to represent a data set. The first method we will consider is *interpolation*. The goal of interpolation is to find a smooth function that fits the data exactly, but the form of the function may be quite complicated and will generally depend on the number of data points in the data set. The other method is *curve fitting*, in which we assume that our function will take a particular (usually simple) form. The function will involve certain parameters and our goal is to determine the values of those parameters that result in the best (approximate) fit between our function and the data.

A.4.1 Interpolation

Suppose we have a data set that consists of N ordered pairs (x_i, y_i), where i is an index that runs from 1 to N. We want to construct a function $f(x)$ such that $f(x_i) = y_i$ for all i from 1 to N. We would like for this function to be "well behaved," which mostly means we want it to be smooth (the function should be continuous and have continuous derivatives up to some order). The hope is that this function will approximate the "true" function that correctly describes not just the data we have but also the data we don't have. We want the function to give us an accurate value for y even when we input a value of x that is not one of our x_i's.

Maxima offers a few options for interpolation, of which we will examine two in this section. The first interpolation method we will consider is *cubic spline interpolation*. Cubic spline interpolation fits the data to a piecewise function. Each piece is a cubic polynomial which is defined on an interval between two consecutive data points. The polynomial pieces are defined so that the function and its first two derivatives are continuous. This ensures that the function passes through all of the data points and also that the function has a smooth appearance (no corners or other sudden changes).

To perform a cubic spline interpolation with *Maxima* we must use the `cspline` command, which is part of the `interpol` package. The code below illustrates how to define a data list, load the `interpol` package, perform the cubic spline interpolation, and assign the resulting interpolation function to a new function that can be called at a later point.

```
(%i) DataList:[[1,3.2],[2.1,7.0],[3.1,12.5],
         [4,11.1],[5.2,6.2],[6,2.3]]$
      load(interpol)$ cspline(DataList)$ c(x):="%;
(%o) (c(x):=
```
$$\left(0.89527\,x^3 - 2.6858\,x^2 + 5.0571\,x - 0.06654\right)\text{charfun2}(x, -\infty, 2.1) +$$
$$\left(0.20749\,x^3 - 3.7348\,x^2 + 17.401\,x - 12.47\right)\text{charfun2}(x, 5.2, \infty) +$$
$$\left(-0.049637\,x^3 + 0.27638\,x^2 - 3.4571\,x + 23.683\right)\text{charfun2}(x, 4, 5.2) +$$
$$\left(2.2047\,x^3 - 26.776\,x^2 + 104.75\,x - 120.6\right)\text{charfun2}(x, 3.1, 4) +$$
$$\left(-3.0755\,x^3 + 22.33\,x^2 - 47.476\,x + 36.706\right)\text{charfun2}(x, 2.1, 3.1)$$

The output function contains several terms. Each term is a cubic polynomial multiplied by a factor of the form `charfun2(x,a,b)`. The function `charfun2(x,a,b)` returns `true` if x is in the interval $[a, b)$ and `false` otherwise. If a `charfun2` factor evaluates to `true` then the polynomial multiplying that factor will be used in evaluating that function. Otherwise the polynomial is not used. Note that the second and third arguments of the `charfun2` functions in the output above form a nonoverlapping set of intervals that cover the real line and have their boundaries at the locations of the data points. Thus, when we evaluate $c(x)$ the function will first determine in which interval x lies and then use the corresponding cubic polynomial for that interval to evaluate the function.

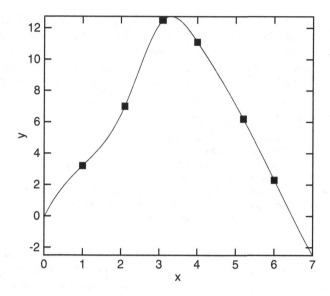

Fig. A.12 The sample data set (*squares*) and the cubic spline interpolating function (*line*)

We can get a sense of how well our cubic spline fits the data by using the now-familiar plotting commands to plot the interpolating function as well as the data points, as displayed in Fig. A.12.

The curve for the interpolating function passes through all of the squares that represent the data points. The function has an overall smooth appearance. We can now use the interpolating function to evaluate y for a value of x within the range of our data (interpolation) or even for a value of x outside of the range of our data (extrapolation).

(%i) [c(2.7), c(7.5)];   (%o) [10.772, −4.5114]

We can expect our interpolating function to give us accurate results (i.e., a good approximation to the "true" function) for values within our data range, provided the true function is well-behaved (smooth, not highly oscillatory, etc.). However, extrapolation is always risky. There is no way to know if our interpolating function is accurate outside of our data range, because we have no way to know what the function is doing outside of that range.

Now that we have our interpolating function, and we feel confident that it is accurate (at least within our data range), we can perform further analysis on that function. For example, we may wish to find the maximum value of y. We can find this maximum value by calculating the derivative $dc(x)/dx$, setting it equal to zero, and solving for x. This gives us the value $x = x_{max}$ that maximizes the function, and then $c(x_{max})$ will give us the maximum value for y (assuming our interpolating function is accurate). However, taking the derivative of our cubic spline function is problematic because we will end up with derivatives of the charfun2 terms which are not defined.

To maximize our cubic spline function we must first determine which piece of the function to use. The plot above shows us that the function is peaked somewhere on the interval $[3.1, 4)$, so we can define a new function $p(x)$ which is just the cubic polynomial from our `cspline` result that corresponds to this particular interval. We can then find the maximum of this polynomial as described above. The output of the code below shows that the interpolating function reaches its maximum value at $x \approx 3.306$ and the maximum value is $y \approx 12.728$.

```
(%i) p(x):=(2.204723215025239*x^3-
     26.77595198132017*x^2+
     104.7521741087083*x-120.5957504953258)$
     diff(p(x),x)$ dp(x):="%$
     matrix(["x at which p is maximized","maximum p"],
     [xmax:find_root(dp(x),x,3.1,4),p(xmax)]);
```

$$(\%o) \quad \begin{bmatrix} \text{x at which p is maximized} & \text{maximum p} \\ 3.306 & 12.728 \end{bmatrix}$$

Another approach to interpolation is to find a single polynomial that fits all of the data points. If we have N data points, then we can fit all of the points by using a polynomial of degree $N - 1$. In fact, a straightforward formula derived by Lagrange defines such a polynomial:

$$L(x) = \frac{(x - x_2)(x - x_3)\dots(x - x_N)}{(x_1 - x_2)(x_1 - x_3)\dots(x_1 - x_N)} y_1 + \frac{(x - x_1)(x - x_3)\dots(x - x_N)}{(x_2 - x_1)(x_2 - x_3)\dots(x_2 - x_N)} y_2$$
$$+ \dots + \frac{(x - x_1)(x - x_2)\dots(x - x_{N-1})}{(x_N - x_1)(x_N - x_2)\dots(x_N - x_{N-1})} y_N.$$
$$(A.11)$$

This function is a polynomial of degree $N - 1$ and $L(x_i) = y_i$ for all i from 1 to N. The `interpol` package includes the command `lagrange` for generating the Lagrange polynomial to fit a given set of data. The code below shows how to use this command to fit our example data set.

```
(%i) fpprintprec:5$ lagrange(DataList)$ L(x):="%$
     expand(%);
(%o) L(x):= -0.14019 x^5 + 2.7237 x^4 - 19.877 x^3+
     65.376 x^2 - 90.534 x + 45.652
```

We can plot the resulting Lagrange polynomial, along with our data points, to check the quality of the fit. The plot is shown in Fig. A.13.

Just as with the cubic spline interpolation, the Lagrange polynomial does pass through each of the data points. However, note the unexpected behavior of the function outside of the data range. In particular, the function shoots up to large values of y as $x \to 0$. This unexpected behavior is a general problem with polynomial interpolation. In fact, we can even get bad behavior *within* the data range if we use a large number of data points to generate the Lagrange polynomial. The more data we have, the higher the degree of the Lagrange polynomial. These high degree Lagrange polynomials may be highly oscillatory even when the data does

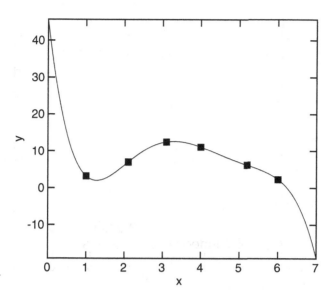

Fig. A.13 The sample data set (*squares*) and the interpolating Lagrange polynomial (*line*)

not indicate that the function should oscillate. For this reason, it is better to perform polynomial interpolation using a small number of data points focused on the range of interest.

For example, to find the maximum value of y within our data range, we might discard the first and last data points (since the peak seems to occur near the center of the data range) and use the middle four data points to compute a Lagrange polynomial. The code below shows how to construct the polynomial and a plot of the polynomial is shown in Fig. A.14.

```
(%i) PeakList:[[2.1,7.0],[3.1,12.5],[4,11.1],
        [5.2,6.2]]$        lagrange(PeakList)$ h(x):="%$
```

Now we can maximize this Lagrange polynomial in order to estimate the maximum of the "true" function.

```
(%i) dh(x):="(diff(h(x),x))$
        matrix(["x at which h(x) is maximized",
        "maximum h(x)"],
        [xmax2:find_root(dh(x),x,2,5),h(xmax2)]);
```

$$(\%o)\quad \begin{bmatrix} \text{x at which h(x) is maximized} & \text{maximum h(x)} \\ 3.2535 & 12.58 \end{bmatrix}$$

We see that this Lagrange interpolation indicates that the maximum value occurs at $x \approx 3.2535$ and that the maximum value is $y \approx 12.58$. These results are similar, but not identical, to the results from our cubic spline interpolation.

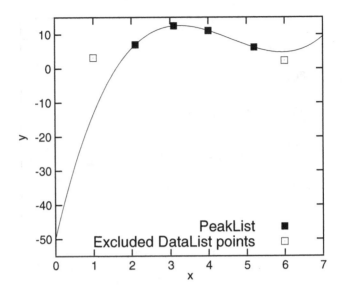

Fig. A.14 The sample data (*squares*) and the Lagrange polynomial (*line*) fit to the middle four data points (*filled squares*)

A.4.2   Curve Fitting

In curve fitting we no longer demand that our model function pass through each data point. Instead, we want to model the data with the equation $y = f(x; a)$, where f is a function of some simple (pre-determined) form and a represents one or more parameters that are used to define the function. We would like to find the parameters values a that result in the function that "best" fits our data.

One problem we must confront is: what do we mean by "best"? What we want is for the function to pass as close as possible to our data points. The best function will be the one such that $f(x_i)$ differs as little as possible from y_i. But we cannot just consider each data point separately. We can force our function to go through any particular data point, but we may then find that the function misses other data points by a wide margin. So what we want to do is to somehow minimize the collective differences between the function and the data points.

There are several ways we can select parameters that minimize this collective difference. The most popular method is to minimize the sum of the squares of the differences between the function and the data points. Thus, we want to find the values for the parameters a that minimize

$$\sum_{i=1}^{N} (y_i - f(x_i; a))^2 .$$ (A.12)

This approach to fitting our function to the data is known as the *least squares method*.

Maxima can perform a least squares fit for a given data set and a given form of the function f. To perform a least squares fit we load the lsquares package and use the lsquares_estimates command. To use this command, we must get our data into the proper format to be accepted by lsquares_estimates. The data must be in the form of a matrix, with each ordered pair (triple, etc.) placed in a row of the matrix. We can convert our sample data list into the proper matrix form using the apply command as shown below.

(%i) **DataMatrix:apply('matrix,DataList);**

$$(\%o) \quad \begin{bmatrix} 1 & 3.2 \\ 2.1 & 7.0 \\ 3.1 & 12.5 \\ 4 & 11.1 \\ 5.2 & 6.2 \\ 6 & 2.3 \end{bmatrix}$$

Now we can load the lsquares package and use the lsquares_estimates command. The lsquares_estimates command takes four arguments. The first is the matrix containing the data, the second is a list of the variables (independent and dependent), the third is an equation specifying the form of the function, and the fourth is a list of the parameters to be adjusted in order to fit the function to the data. The code below shows how to fit our sample data to a quadratic function $g(x) = Ax^2 + Bx + C$.

(%i) **load(lsquares)$**
     params:lsquares_estimates(DataMatrix,
       [x,y],y=A*x^2+B*x+C, [A,B,C]);
     float(params);
(%o) $[[A = -\frac{240273530}{167471391}, B = \frac{1660389386}{167471391}, C = -\frac{6497960877}{1116475940}]]$
(%o) $[[A = -1.4347, B = 9.9145, C = -5.8201]]$

Note that by default *Maxima* gives the parameter values as exact values if it can. These values can easily be converted to decimal form using float.

We can examine the fit by plotting our quadratic function with the parameter values determined by the least squares fit along with the data points, as shown in Fig. A.15.

The fit is not bad, but obviously the curve does not pass through all of the data points (or even though one of the data points). Even so, it appears plausible that this is the best quadratic function to fit the data. The peak seems to be in the right place. In fact, it is not hard to show that the function must be peaked at $x = -B/(2A)$. With the parameter values from our least squares fit this indicates that the peak is at $x \approx 3.455$, which is similar to the results from the interpolating functions discussed above. However, the peak value for our quadratic function is $y \approx 11.308$, which is noticeably less than the values obtained from the interpolating functions.

It is not surprising that this simple quadratic function cannot fit the data as well as our interpolating functions did. Recall that the cubic spline function consisted of five pieces, each of which was a cubic polynomial. The Lagrange polynomial was a polynomial of degree five. Therefore, the interpolating functions have many more

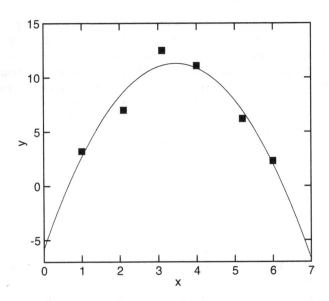

Fig. A.15 The sample data set (*squares*) and the quadratic least squares fit function (*line*)

parameters that can be tweaked to give a better fit. Depending on your goal, you may prefer a simple function that gives a decent fit to the data to a very complicated function that gives a perfect fit.

A.5   Exercises

1. Use the bisection method to find the roots of the function $f(x) = \sin(x) - x^2 + \log(x) + 2$.

   (a) Plot the function on the interval $0 < x \leq 4$. Determine the approximate location of the two roots of this function on this interval.

   (b) Suppose we were to apply the bisection algorithm to finding the root of $f(x)$ near $x = 2$. Suppose we set our initial interval to $[1.5, 2.5]$ so that it contains that root and no other roots. Using the terminology of the text, we are starting with $x_L = 1.5$ and $x_R = 2.5$. Complete the table below to determine the values of x_L, x_R, and $x_M = (x_L + x_R)/2$ before each pass through the loop in the bisection algorithm. Also show whether the function $f(x)$ evaluated at these x-values is positive $(+)$ or negative $(-)$. Finally, show whether $x_M \to x_L$ or $x_M \to x_R$ for the next pass through the loop.

i	Before			Function Values		After	
	x_L	x_R	x_M	$f(x_L)$	$f(x_R)$	$f(x_M)$	$x_M \to$?
1							
2							
3							

(c) Use the bisection algorithm to find the two roots of $f(x)$ using an error tolerance of 10^{-6}. Record your starting values for x_L and x_R, along with the root (x_r) and the number of passes through the loop, in the table below.

x_L	x_R	x_r	# Passes

(d) Use find_root to check your values for the roots of $f(x)$.

2. Consider the function

$$f(x) = e^{-x^2}\sqrt{x^2 - 1}.$$ (A.13)

(a) Use integrate to find an exact expression for the integral

$$\int_1^3 f(x)\, dx.$$ (A.14)

Explain why the result is not particularly useful.

(b) Use the trapezoidal approximation to evaluate the above integral. Play around with the number of divisions n until you are sure your answer is correct to three decimal places. What is the minimum number of divisions you have to use to obtain this precision (i.e., what is the smallest value of n you can use to get a result that is accurate when rounded to 3 decimal places)?

(c) Use Simpson's Rule to evaluate the above integral. Make sure your answer is precise to three decimal places. What is the minimum number of divisions you have to use to obtain this precision? Does this result agree with what you got from the trapezoidal approximation, at least to three decimal places? Would you say that Simpson's Rule is more efficient, less efficient, or about the same efficiency as the trapezoidal approximation for evaluating this integral?

(d) Use one of *Maxima*'s built-in routines to evaluate the above integral. Does the result agree with your values from the trapezoidal approximation and Simpson's Rule to at least three decimal places?

3. Suppose we wanted to numerically estimate the value of

$$\int_0^2 \int_0^5 \int_0^{10} q(x, y, z) \, dz \, dy \, dx, \qquad\qquad \text{(A.15)}$$

where

$$q(x, y, z) = x^2 yz^3 e^{-x^2-y^2-z^2}. \qquad\qquad \text{(A.16)}$$

(a) Which numerical integration method would you use, and why?
(b) Use `integrate` to determine the exact value of this integral, then convert to a floating-point number.
(c) Use the numerical method you specified to estimate the value of this integral. Record your results in the table below.

n	Estimate	Correct Sig Figs
10		
100		
1000		
10,000		
100,000		

Comment on your results.
(d) Set `rombergtol` to 10^{-6} and then use `romberg` to compute this integral. To how many decimal places is the result accurate?

4. Consider the quartic oscillator system, which consists of a particle subject to the force $F = -4\alpha x^3$ (see Problem 13 in Sect. 5.5).

(a) Use the built-in `rk` command to solve the equations of motion for the quartic oscillator with $m = 1$ g, $\alpha = 1$ erg/cm^4, $x(0) = 1$ cm, and $v(0) = 0$. Use a time step of 0.1 s and integrate the equations of motion from $t = 0$ to $t = 10$ s. Construct plots of position versus time, velocity versus time, and the trajectory in phase space (velocity versus position). If you completed Problem 13 from Sect. 5.5, discuss any differences you see between these results and those from the Euler–Cromer algorithm.
(b) Construct a plot of energy versus time for the quartic oscillator, using your 4th order Runge–Kutta algorithm (the algorithm used by `rk`) results. Does this algorithm conserve energy (on average)? You know that total energy should be constant—what value should it have? What is the maximum relative error (i.e. the difference from the correct value divided by the correct value) in the energy produced by the `rk` command over this time interval?

5. Consider a particle of mass $m = 0.1$ kg that moves along the x-axis in a potential well of the form $V(x) = -k/(x^2 + a^2)$, with $a = 0.1$ m and $k = 0.05$ J·m^2.

(a) What is the force on this particle? Construct plots of both the potential energy function and the force as a function of x.

(b) Write down Newton's Second Law for this system as a system of two first-order ODEs.

(c) Use the built-in `rkf45` command to solve the equations of motion for this system with $x(0) = 0.2$ m, and $v(0) = 0$. Use a time step of 0.01 s and generate solutions for $0 \leq t \leq 1$ s. Construct plots of x versus t, v versus t, and v versus x. Comment on how this motion compares to that of a harmonic oscillator.

(d) Use your results to construct a plot of energy versus time. Is the energy conserved, on average? You know that total energy should be constant—what value should it have? What is the maximum relative error (i.e. the difference from the correct value divided by the correct value) in the energy produced by the `rkf45` command?

6. Consider the following data set of (x, y) ordered pairs: $[[6.9, 61.5], [9.1, 63.7], [11.3, 68.0], [13.5, 65.4], [15.7, 58.9], [17.9, 50.6]]$.

(a) Use `cspline` to create an interpolating function for this data. Find the value of x that maximizes the interpolating function, as well as the maximum value of the interpolating function.

(b) Use `lagrange` to create an interpolating function for this data. Find the value of x that maximizes the interpolating function, as well as the maximum value of the interpolating function. Compare to your results from `cspline`.

(c) Fit this data to a quadratic polynomial $y = Ax^2 + Bx + C$. Find the value of x that maximizes the fit function, as well as the maximum value of the fit function. Compare to your results from `cspline` and `lagrange`.

7. Show that a least squares fit to the sample data in Sect. A.4 for a fifth degree polynomial $h(x) = Ax^5 + Bx^4 + Cx^3 + Dx^2 + Ex + F$ gives the same function as the `lagrange` interpolation method. Explain, with reference to the number of parameters and number of data points, why this should be the case.

Index

Printed in the United States
By Bookmasters